A Platform-Centric Approach to System-on-Chip (SOC) Design

Vijay K. Madisetti
Chonlameth Arpikanondt

A Platform-Centric Approach to System-on-Chip (SOC) Design

Springer

Cover images of modern military technology courtesy of U.S. Department of Defense

Library of Congress Cataloging-in-Publication Data

A C.I.P. Catalogue record for this book is available
from the Library of Congress.

ISBN 0-387-23895-6 e-ISBN 0-387-23896-4 Printed on acid-free paper.

Printed in the United States of America.

9 8 7 6 5 4 3 2 1 SPIN 11352082

Dedication

VKM dedicates this book to Anitha and Raj

CA dedicates this book to his family – mom, sis, and app.

Contents

Preface

Increasing system complexity has created a pressing need for better design tools and associated methodologies and languages for meeting the stringent time to market and cost constraints. Platform-centric and platform-based system-on-chip (SoC) design methodologies, based on reuse of software and hardware functionality, has also gained increasing exposure and usage within the Electronic System-Level (ESL) design communities.

The book proposes a new methodology for realizing platform-centric design of complex systems, and presents a detailed plan for its implementation. The proposed plan allows component vendors, system integrators and product developers to collaborate effectively and efficiently to create complex products within budget and schedule constraints. This book focuses more on the use of platforms in the design of products, and not on the design of platforms themselves.

Platform-centric design is not for everyone, as some may feel that it does not allow them to differentiate their offering from competitors to a significant degree. However, its proponents may claim that the time-to-market and cost advantages of platform-centric design more than compensate for any drawbacks.

We are grateful to the faculty, students, and industry partners, participating in the State of Georgia's Yamacraw Embedded Software Research Program at the Georgia Electronic Design Center (GEDC) of Georgia Institute of Technology for their support of our research in the past four years. We also thank professors Roger Webb, Nikil Jayant, Rao Tummala, Ron Schafer, Joy Laskar, and Herb Lehman at Georgia Tech, and Venu Dasigi of Southern Polytechnic State University (SPSU) for their encouragement and support.

Acknowledgments

This research was supported in part by the State of Georgia's Yamacraw Research Program, in part by US Army's Federated Research Laboratories Program on Sensors, and in part by DARPA's Rapid Prototyping of Application-Specific Signal Processors (RASSP) program between 1996 and 2003 at the Georgia Electronic Design Center of the Georgia Institute of Technology, Atlanta, GA 30332, USA. Validation of the methodology proposed in the book formed the basis of Dr. Arpnikanondt's Ph.D. dissertation at Georgia Institute of Technology in 2004.

We thank Mr. Brian Bell of Soft Networks, LLC, and Ms. Rebecca Olson, and Mr. Michael Hackett of Springer for their assistance with the book.

Chapter 1

INTRODUCTION TO SOC DESIGN

1. INTRODUCTION

The semiconductor industry is doing well, with worldwide sales in the year 2003 grossing an approximate 180 billion USD—30 percent of which came from microprocessors, DSPs, microcontrollers, and programmable peripheral chips [1]. A recent forecast by Dataquest predicts an estimated 200+ billion USD in semiconductor sales by the end of the year 2004 [130]. Along with these huge market opportunities, however, has come an increase in system design complexity. It is estimated that by the year 2010 the expected transistor count for typical System-on-Chip (SoC) solutions will approach 3 billion, with corresponding expected clock speeds of over 100 GHz, and transistor densities reaching 660 million transistors/cm2 [2]. Concurrently, this increase in complexity will result in an increase in power dissipation, cost, and the "design-to-market" time.

In [2] it is argued that computer products will eventually progress from large, general-purpose, impersonal static forms to portable, personal, flexible, market-targeted forms. Personalization, flexibility, and quick time to market will dictate a "quickturn" design approach. Time-to-market for new platforms will no longer be measured in years, but in months or weeks. Design cycle must decrease or become the bottleneck for future progress and may determine which corporation survives and which does not. The stringent time-to-market requirement, coupled with additional design constraints such as design flexibility, cost, real-time requirements and rigid form factor (e.g. size, weight and power dissipation) represent a formidable challenge that designers of current systems must overcome. Today, the need for a paradigm shift in the design method has become more and more pronounced and demanding.

1.1 TYPICAL SYSTEMS

An embedded system-on-a-chip (SoC) platform, with programmable or dedicated logic and multiple functional units sharing a common memory, similar to that shown in Figure 1.1, is gaining popularity rapidly and may become the option of choice among system-level houses [3]. This assertive conclusion stems from many different factors, some of the more important ones being listed below.

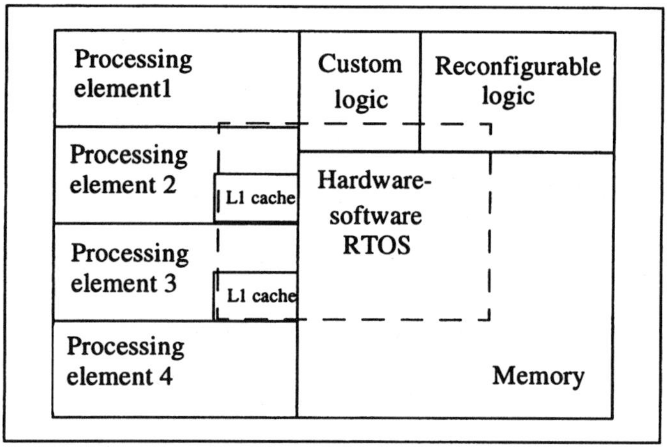

Figure 1-1. Typical system-on-a-chip (SoC) architecture

- In terms of sales volume, embedded processors have outsold PC processors by a large margin [1]. A wide range of applications, especially in the area of wireless and portable devices, has attributed to a tremendous demand for embedded processors. Given a rate of progress in the IC technology today, a real-time embedded SoC will find even more suitable applications in the future, with IP reuse featured prominently (Figure 1-2) as a method for improving productivity.

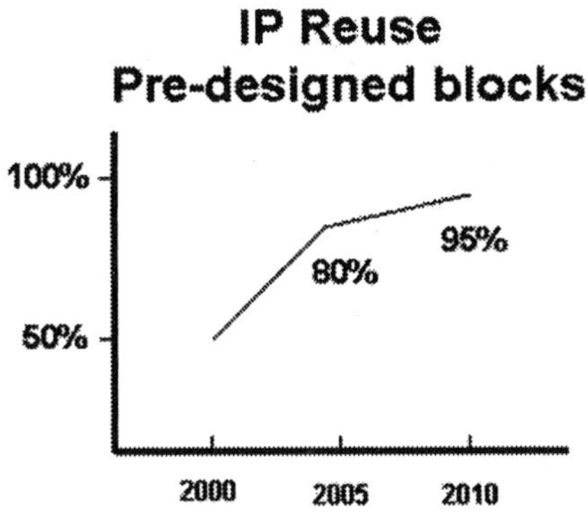

Figure 1-2. Expected Use of Pre-Design Components in New SoC Designs

- Current power density data [4] show that power tends to double every eighteen months. Herring [2] discusses that power dissipation will have a major role in determining what systems in the future would look like. The power density extrapolation as depicted in Figure 1.2 shows that the dissipated heat will eventually approach those of nuclear power (250 watts/cm2)!. With such an increase in power dissipation, the costs associated with packaging and thermal dissipation may dwarf any savings achieved with higher transistor densities. Noise and coupling issues will also become even more difficult to solve as frequencies increase. As a result a distributed system that contains several smaller, slower, heat-manageable processors is going to become a preferable choice to a powerful single-processor system with a hard-to-solve power dissipation problem.

- As the IC technology progresses from micron to submicron to molecular levels, the cost of producing mask sets will skyrocket [31]. A new 300nm, 0.13μ or 0.10μ high-volume manufacturing plant today cost about $3.5 billion [32]. As such, chip designers will tend to produce a chip suitable for more than one application. Coupled with the fact that performance flexibility and application flexibility are so essential to the success of an embedded system [2, 5], a system with a high degree of reconfigurability may not be uncommon in the near future.

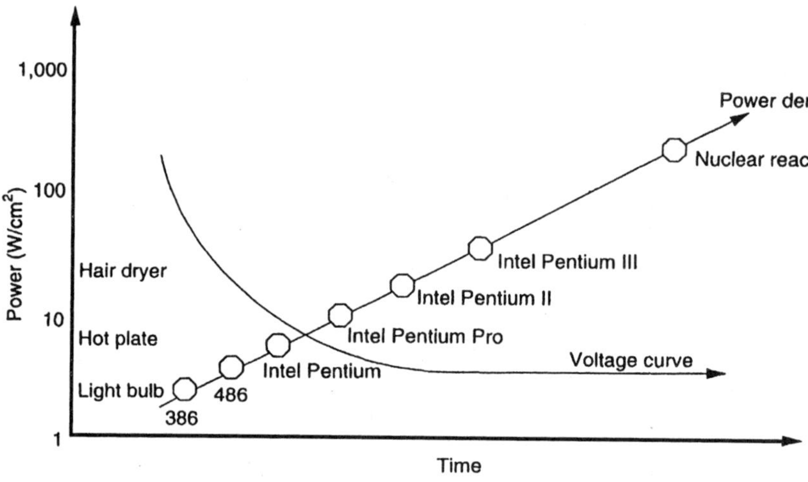

Figure 1-3. Power density curve

This book will focus upon a design method that is more suitable for such systems and their predominant requirements such as minimal time-to-market, real-time environment, flexibility and various form-factor constraints. A paradigm shift in current co-design approaches is imperative. Such design approaches that derive from a processor running sequential code have fast become a legacy and uncharacteristically tedious when dealing with the development of systems today. An ever-growing system complexity and pressure to keep the technology-to-market time to the minimum further dictate the need for an improvement in current co-design practices.

1.2 TRADITIONAL CO-DESIGN METHODS

Hardware-software co-design refers to a concurrent and cooperative design of hardware and software in a system. As shown in Figure 1.3, a process of co-design usually involves four main tasks: architecture selection, HW/SW partitioning, scheduling and communication synthesis. The design process flows in a waterfall style, often commencing with a set of specifications. This approach, however, suffers from numerous limitations:

- Use of written requirements
- Lack of a collaborative hardware-software co-design environment

- Limited architecture exploration
- Lack of distributed, real-time support

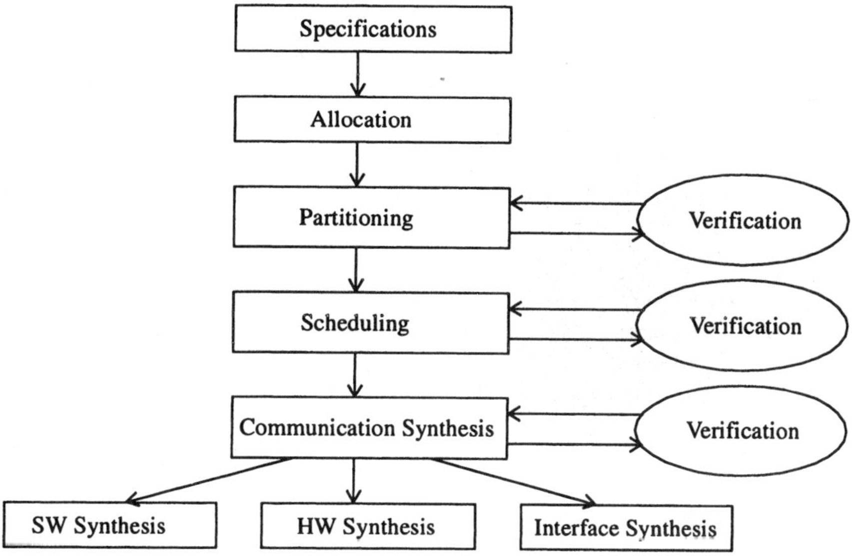

Figure 1-4. Generic hardware/software co-design process flow

- Inability to cope with very complex SoC systems.

Using written requirements to specify system functions and constraints fosters ambiguity in interpretation and does not facilitate customer interaction, thus leading to increased design iterations and low customer satisfaction. Current industrial practice commonly relies upon designer experience and ad hoc techniques to select an architecture and allocate algorithm functionality [6]. This approach severely limits a designer's ability to explore the design space to improve the design.

To minimize overall system costs, traditional system-level design and test approaches attempt to minimize hardware costs, subject to performance constraints. However, these approaches overlook an important characteristic of software prototyping. Parametric studies based on historical project data show that designing and testing software is difficult if margins of slack for hardware CPU and memory resources are too restrictive [6]. Figure 1.4 [7] depicts the software prototyping principle. Graphs (a) and (b) show that software dominates system development cost and time when CPU and memory utilization are high. However as developers reduce resource utilization by adding extra hardware resources software cost and schedule tend to decrease drastically. SoC and distributed-processing chip-building techniques are, historically, not the cause of production delays. Software availability, support and knowledge base are the bane of product schedules [2].

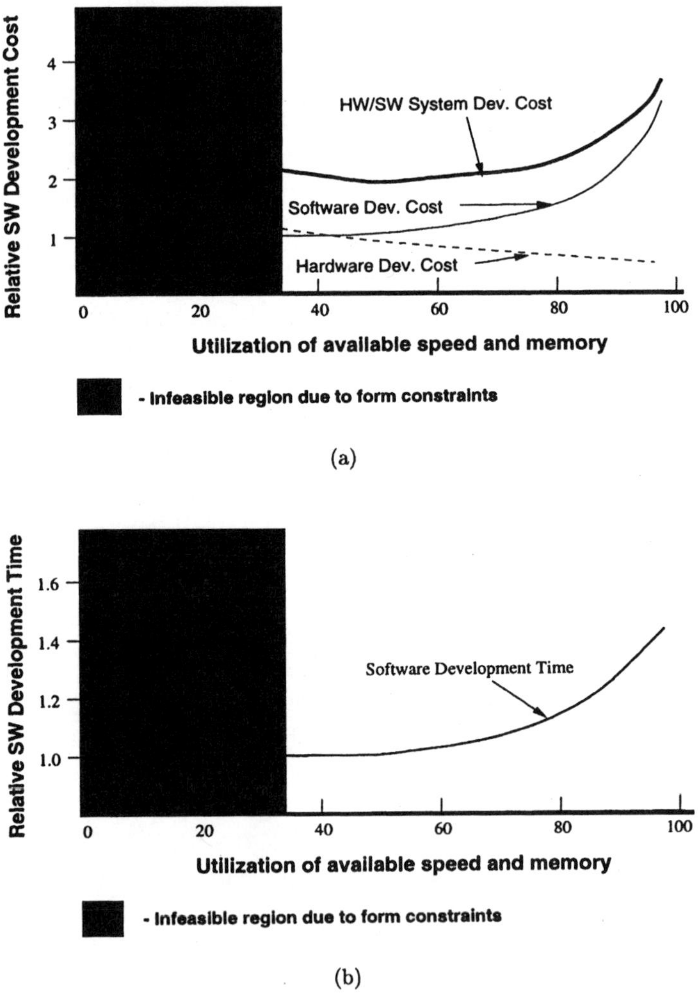

Figure 1-5. The effect of hardware constraints on: (a) HW/SW prototyping costs, and (b) software schedule [Madisetti].

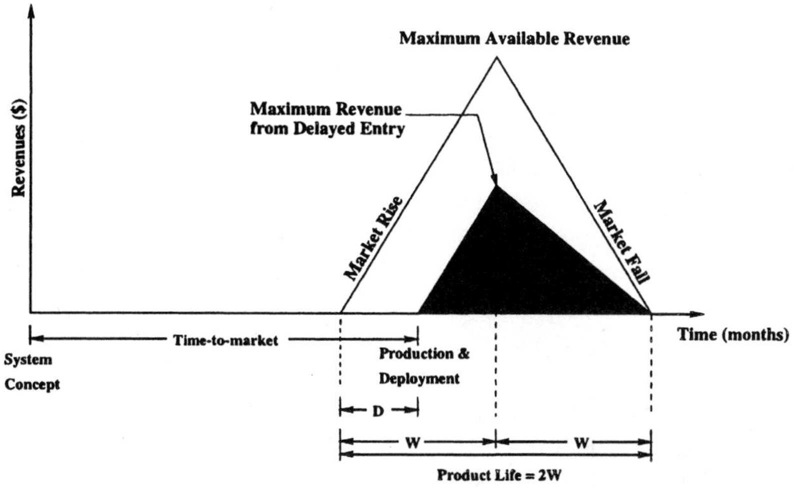

Figure 1-6. A typical time-to-market cost model.

While constraining the HW/SW architecture is detrimental to software development cost, the corresponding effect on development time can be even more devastating. Time-to-market costs can often outweigh design, prototyping, and production costs of commercial products. Figure 1.5 illustrates a model developed by ATEQ Corporation [8] to study the economic impact of delayed market introduction. This simple model quantifies lost revenue from delayed market entry (d) based on the product's total projected revenue and the duration of the market window (W). The unshaded region of the triangular revenue curve depicts this revenue loss. When the product life cycle is short, being late to the market often means a high risk of financial loss resulting in a crisis.

Friedrich et.al., [5] report that, in most embedded applications, the use of general-purpose operating system platforms is not applicable because they are too expensive. Embedded system requirements such as processor performance, memory, and cost are so variable that a general-purpose operating system may not meet all the needs of customers. Other approaches, such as avoiding an operating system altogether by implementing all the functionalities directly, or by developing an in-house operating system, may limit flexibility and be costly. However, a survey suggests that general purpose platforms are used by 66 percent of the embedded systems in Japan, primarily because a suitable alternative does not readily exist. As such, providing a real-time operating system support with a high degree of reconfigurability becomes very essential to the development of systems today.

Most current HW/SW co-design methods commence with a set of specifications that is usually described by some kind of a formal language. However, as the design process goes from the gate to register-transfer level (RTL), from RTL to the instruction set, and from the instruction set to system level design where formalism often does not exist, problems start to ensue [9]. Furthermore traditional HW/SW co-design approaches still rely heavily on a simulation-based technique, where designers go top-down until getting the design, then describing it in some language, simulating it, and figuring out what needs to be done. With such an approach, decisions made during the design process are usually based solely on designers' experiences. Also traditional co-design practices require more time to correct any constraint violation because these constraints often are checked very late in the design process. Given the complexity and demanding schedules of today's commercial systems, these blind, design-first-constraint-checked-last approaches tend to yield a sub-optimal result.

1.3 TOOLS INTEGRATED ENVIRONMENT

As traditional HW/SW co-design methods fail to address many issues involved in the development of SoC systems today, the first generation of integration tools have emerged that attempts to aid systems developers in coping with an increasing system complexity. Such tools normally run on a single underlying semantic backbone, or a single model of computation [9].

In a tools-integrated environment, the design is usually captured into some kind of a unified representation, e.g., Specification and Description Language (SDL), or Co-design Finite State Machine (CFSM). During the design flow, which virtually still adheres to that of the traditional co-design approach, this unified representation acts as an input to a collection of different tools, and often change during each design stage to incorporate more and/or new information into the design. The Corsair design flow [10] illustrated in Figure 1.6 is a good representation of this co-design approach. Other tools-integrated environments include such methods as POLIS [11] and CoWare from Cadence Design Systems.

Although the "tools integration" method may offer designers a better automated and integrated HW/SW co-design environment, its tendency to adhere too much to the traditional co-design practice may be the downside. Its aim to tackle the complexity of today's SoC systems may eventually find a bottleneck in its ineffectiveness to raise a level of design abstraction that is only acceptable to the traditional co-design method. Also, like its traditional precursor, the lack of focus on flexibility can be a disadvantage for current SoC systems that are so application-oriented and highly driven by technology-to-market time.

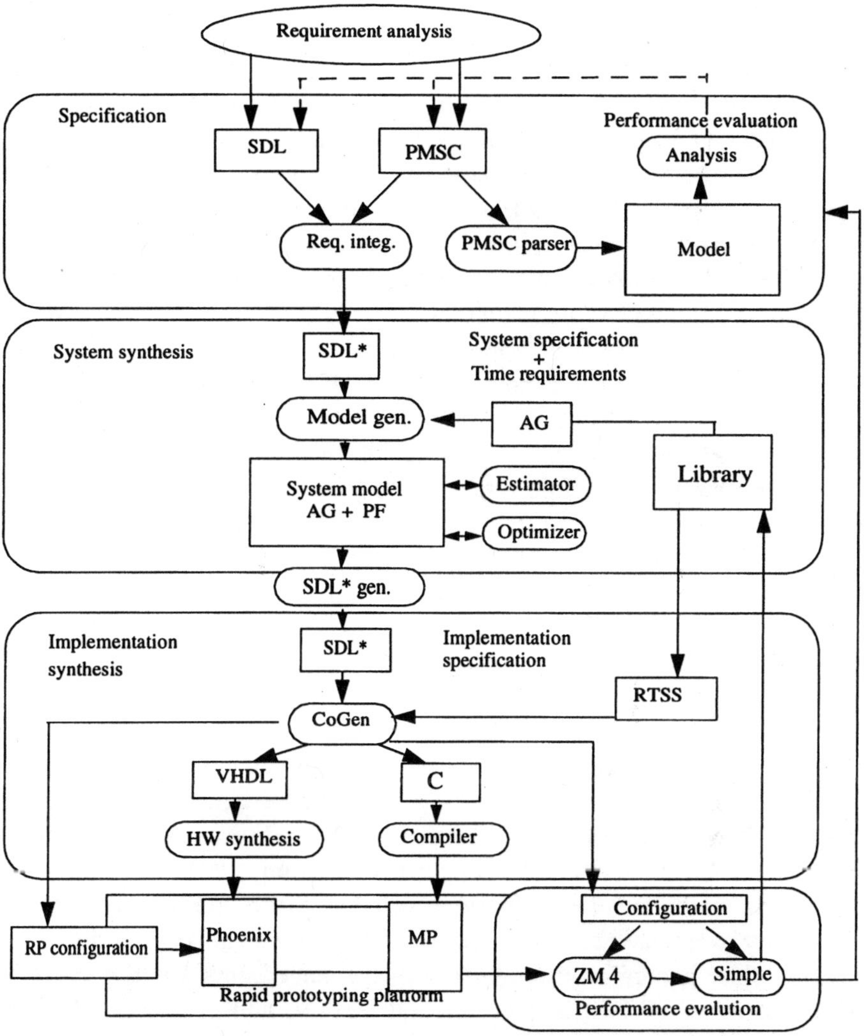

Figure 1-7. The Corsair design flow.

1.4 PROBLEM STATEMENT

It appears, from the previous discussion, that most existing HW/SW co-design methods inadequately address the issues involved in the development of real-time, embedded SoC systems. As the system complexity grows in accordance with Moore's law, and as a pressure to minimize the technology-to-market time increasingly overwhelms, the necessity for a paradigm shift in HW/SW co-design practices becomes a mandate [2]. Flexibility dictates that a processor be designed with the absolute highest general-purpose performance possible [9]. More and more systems developers will turn to build a system out of such processors to meet tight time-to-market constraints and flexible application requirements [2]. Constraints must be fully addressed to ensure reliability. In addition, as semiconductor manufacturers continue to define new methods and new ways to build systems, it is desirable for systems developers to be able to incorporate such technological advances into their existing design approaches. Nevertheless, the fact stands that the semiconductor market is too vast and too dynamic for current HW/SW co-design approaches to readily keep pace with its demands. It is also an extremely competitive market, where such a slack often proves costly to systems manufacturers.

Motivated by all of the reasons above, this book attempts to improve upon the traditional co-design method and the tools-integrated approach. It aims to raise a design abstraction level as well as to improve the various vital aspects essential to the success of a "quickturn", yet reliable, development of SoC systems [138].

1.4.1 Technical Problem

As discussed in DeBardelaben and Madisetti [6], a combinatorially significant number of alternatives exist in the implementation of embedded systems. Compounded by increasing complexity, the problem of implementing such systems becomes even more difficult. Embedded system requirements are also invariably diverse and often conflicting. Some key requirements include such attributes as battery life, portability, security, connectivity, user interface, application compatibility, universal data access, and cost. This list of requirements presents an enigma for the CPU and system developer. While battery life, portability, and cost require simple, application-specific solutions, universal data, security, and user interfaces require the ability for higher performance. These requirements call not only for specific digital performance requirements, but also for specialized analog capability to permit better interaction with the analog-centric human user. CPU and system optimization for one set of requirements will cause unacceptable design trade-offs in other areas. For example, architecture cannot be designed solely for high-overhead, general-purpose performance, or it will sacrifice battery life, portability, and cost.

The objective of this book is, therefore, to develop a systematic approach and guideline that can be used as a design framework to assist system developers in:

- carrying out the SoC design with quickness and correctness,
- exploring architecture and design space so that optimal decisions can be made, and,
- coping with an ever-growing complexity of systems.

The efficiency of this approach is to be comparatively evaluated using the Constructive Cost Modeling technique, COCOMO II.2000 [19].

1.4.2 Technical Challenges

Designing a real-time embedded SoC is a difficult task in its own right as there are often many conflicting requirements to be reckoned with. To support system developers, however, with an efficient co-design method that aids a wide spectrum of such design can prove even more difficult. Some of the more formidable challenges are listed as follows:

- Determining a suitable approach to facilitate the system developer in dealing with an invariably diverse set of requirements as well as the increasing complexity, while improving upon the "technology-to-market" time,
- Developing a design environment that fosters the use of a wide variety of state-of-the-art design tools and techniques, and is adaptive to frequent technological changes,
- Developing a method that has a good appeal towards semiconductor vendors, EDA tool vendors and system developers.

The complexity of these tasks are primarily attributed to the following factors:

- Large size of the design space: A combinatorially significant number of architectural and functional alternatives exist in the implementation of embedded SoC systems. Available components often vary in cost, performance, modifiability, reliability, power, size, and design effort. In addition, there are a lot of communication elements (buses, crossbars), communication protocols (Bluetooth, PCI), and interconnect topologies (ring, linear, mesh, tree) to choose from. Further compounded by various combinations of requirements, the design space that system developers must explore, becomes enormously expansive.

- Complexity and constraints imposed on design time and cost: The year 2002 has seen information appliances outsold PCs by a wide margin [1]. This new market encompasses small, mobile, and ergonomic devices that provide information, entertainment, and communications capabilities to consumer electronics, industrial automation, retail automation, and medical markets. These devices require complex electronic design and system integration delivered in the short time frames of consumer electronics. The system design challenge of at least the next decade is the dramatic expansion of this spectrum of diversity and the shorter and shorter time-to-market window [13].

- Tools and techniques: Vissers [9] argues that semiconductor manufacturers will design systems, rather than system houses design processors. The semiconductor sector is going to do more co-design, with a lot of knowledge that was previously at the system house being either handed off to, or moving towards the semiconductor company. As such, semiconductor manufacturers will tend to define new methods and new ways to build systems, and the tools for system design will need to be based on the same design approach. Given such a trend, the design method must be able to provide a unified environment for tools and techniques from both the semiconductor and the system houses.

- Necessity for a paradigm shift: Traditional co-design approaches are no longer a viable choice for handling the complexity of today's embedded SoC systems. For example, system developers using traditional co-design approach often make design decisions blindly and a priori for there is a lack in the availability of pre-simulated data and/or cost-estimation tools. Techniques such as standardization, co-simulation, co-verification, co-synthesis, reuse, etc. need a refreshing re-examination. Systems complexity dictates that a design abstraction level be raised higher. The need for a paradigm shift in co-design approaches becomes eminent. A new method, to be useful, must satisfactorily address these requirements associated with current embedded SoC systems.

- Disagreement on a standard practice: While adhering to a standard design practice and/or a standard set of tools actually helps, it is extremely unlikely that there will ever be a unanimous agreement on any one design standard. Marketing strategies, legacy designs and many other opposing factors are often weighed in heavily. Consequently, many excellent standard and non-standard design approaches will continue to co-exist. To make use of them to the fullest, the design method must act like a system of such tools and techniques that allows them to work in unison within the same environment. This integration effort will also prove to be difficult.

1.4.3 A Solution – The Codesign Modeling Framework (CMF)

To effectively address the issues in co-design, developers may have to raise a design abstraction level so as to better foster extensive reuse and make a better use of cutting edge tools and techniques. In addition, requirements must be systematically taken into account to better reflect the various needs and constraints that are pertinent to the systems development. This book presents a design approach, called a "platform-centric" SoC design method, or Codesign Modeling Framework, and is largely motivated by the systems development model as depicted in Figure 1.7.

The development model in Figure 1.7 views the systems development environment as consisting of three separate domains: a reusable object domain, a platform domain, and a product design domain. The reusable object domain provides a library (or libraries) of logical platform-compliant tool and component entities, whose physical whereabouts could virtually be anywhere accessible through their logical counterparts in the library. Such use of a logical library within the reusable object domain allows various forms of platform-compliant tools and components to be uniformly abstracted and represented, and pre-characterized data compactly modeled and efficiently utilized in the platform and the product design domains.

Chapter 4 of this book details such a logical library, called a library of platform objects (LPO), that constitutes the reusable object domain. This book utilizes the Extensible Markup Language (XML) to model the logical LPO database. XML permits the power of many existing Internet technologies to be harnessed that could potentially lead to an efficient exchange of data—be it knowledge, application programs, or design components.

When carefully designed, it is also possible for a reusable platform object and/or application program to insert itself into a reusable object domain as an LPO module, thus making the library scalable and as expansive as the Internet itself.

On the other hand, the platform concept [96] helps expedite the design process by reducing the degree of freedom during the development process, without absolutely relinquishing system flexibility and performance. The platform domain comprises platforms that are tailored for various specific purposes, e.g., workstation, low-power handset, VDO, or multimedia. For each platform, the platform vendor would normally also specify and/or supply compatible tools and components in the reusable object domain that can be used to develop the final product.

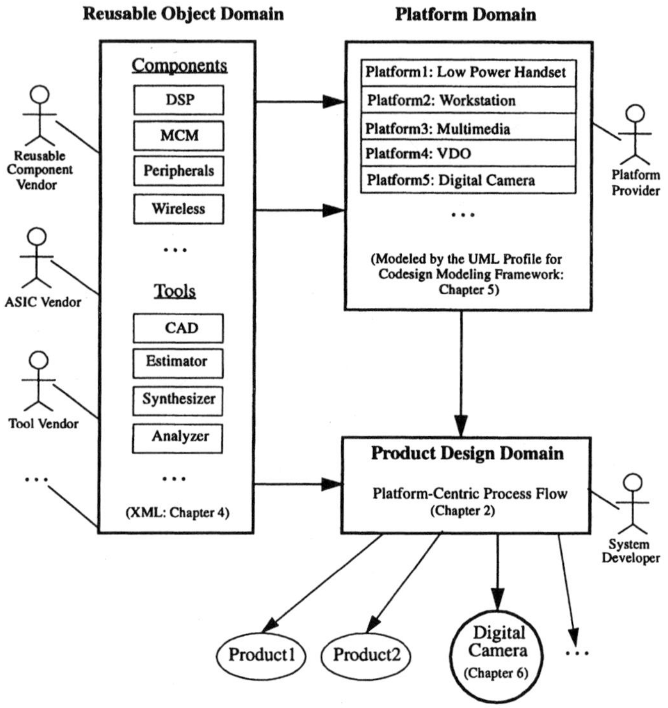

Figure 1-8. The enhanced system development mode, Codesign Modeling Framework, as proposed in this book.

The system development phase that renders the final products takes place in the product design domain. Within this domain, the system developer employs the platform in the platform domain and the platform-compliant tools and components in the reusable object domain to derive the target architecture—and eventually the final product. The product design process is described in detail in Chapter 2; it is then applied to develop a simplified digital camera system as an application case study in Chapter 6.

The Unified Modeling Language (UML) plays an integral role. By itself, UML furnishes modeling capabilities for handling real-time requirements in software systems. Chapter 5 of this book further enhances such capabilities through UML's stereotypes and tags such that, by using the proposed approach, (1) HW design and HW/SW co-design are robust and possible, (2) platforms could be modeled efficiently, (3) the system developer could conveniently model interrupts and exceptions—the characteristics that are vital to most real-time embedded systems, and (4) the enhanced UML for the proposed approach would furnish an interface facility to the reusable object domain, as well as the product design domain. The extended UML elements provided for by this book are packaged together in the UML Profile for Co-design Modeling Framework as depicted in Figure 1.8, and detailed in Chapter 5.

The platform-centric SoC method is aimed at the design of today's SoC systems with emphasis on real-time, embedded systems. The approach provides a guideline and a SoC design environment that promotes an integration of state-of-the-art tools and techniques necessary for the development of the systems. It renders a new and better perspective towards co-design approaches, while also raising a level of design abstraction. Because the configurable platform objects are designed off-cycle, they contribute to a general improvement in development time. By incorporating their usage, the overall method strikes a balance between total design flexibility and minimal time-to-market.

Within a platform-centric environment, timing behaviors and other constraints (e.g. size, weight) are more predictable. Detailed functional specification derived in the analysis phase will be mapped directly to the target architecture, which, in turn, is constructed using platform compatible hardware and software components, much like how a personal computer system is built by selecting from a menu of different available options. At the core of the platform-centric approach lie the Unified Modeling Language (UML) and a library of platform objects (LPO) that, together, effectively raises the design abstraction level and promote faster development time without incurring additional costs. The approach also permits a seamless integration of tools and techniques that aid analysis and synthesis processes as well as design automation.

Borrowed from the software engineering field, and adapted to better suit the requirements for the development of real-time embedded SoC systems, UML represents a force that drives the development process flow right from where system requirements are analyzed and captured until the desirable implementation model results. The use of UML as a unified representation of the system under development eases the constraints processing and requirements analyzing processes. Its object-orientedness allows complexity to be effectively handled. UML uses its own constraint capturing mechanism and the supplementary Object Constraint Language (OCL) in dealing with a wide variety of constraints. This, in effect, along with the UML modeling capability serve as the underpinnings for the derivation of the platform-independent functional specifications with timing requirements. The UML profile for schedulability, performance, and time specification [29], while has yet to be fully standardized, is useful for modeling real-time systems for analysis and synthesis purposes. The semi-formal nature of UML practically bridges the analysis-synthesis gap to yield a better design flow. UML framework for hardware and software unified modeling of real-time, embedded systems will be presented.

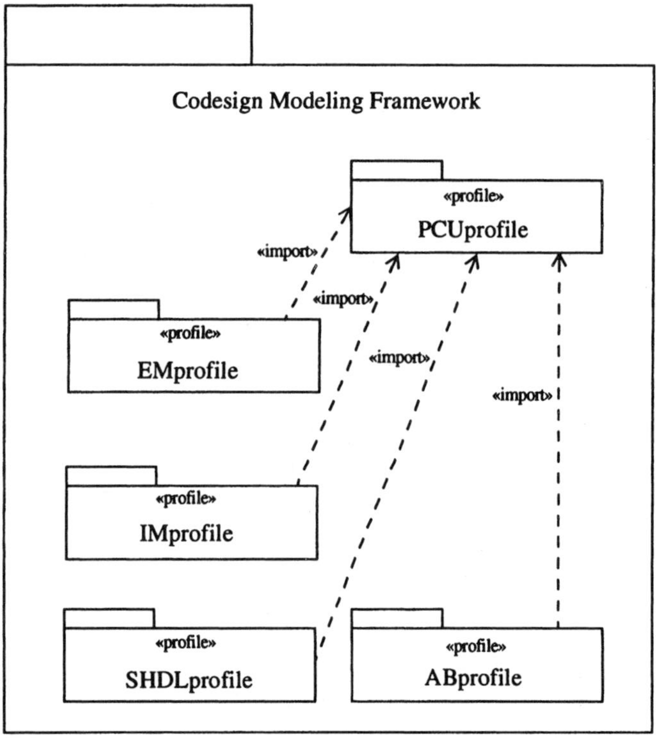

Figure 1-9. The UML Profile for Co-design Modeling Framework (see Chapter 5).

The library of platform objects (LPO), on the other hand, provides a collection of system platforms, i.e. platform objects (PO), that are further governed by a set of rules and requirements specific to the proposed platform-centric SoC design method. Each platform object represents a collection of a common configurable architecture along with its related components, and tools and techniques specific to that platform. The eXtensible Markup Language (XML) can be used to provide a concrete facility for realizing the LPO and managing the complexity resulted from a huge database of extremely diverse LPO modules. XML also promotes information interchange which can culminate to the virtually open-source library of platform object member modules. Furthermore, there exists an extensive collection of public-domain and/or open-source tools for XML and other related technologies that can be used to enhance XML's capability.

1.5 ORGANIZATION OF BOOK

In this chapter, some challenges faced by developers in the co-design of SoCs have been described. The remainder of this book presents a more thorough examination of the problem and the proposed approach to a solution.

Chapter 2 describes the proposed platform-centric SoC design method in detail. It illustrates the design flow and discusses each step in the design process. The definition of a platform as originally defined by Sabbagh [96], as well as recent platform-based and platform-centric design approaches, are presented. The chapter concludes by comparing the proposed approach with current literature on design methodologies.

Chapter 3 covers some of the enabling technologies behind the proposed SoC design methodology. Whereas the platform technology is discussed in Chapter 2, this chapter gives an overview of the other two fundamental technologies: the Unified Modeling Language (UML) and the Extensible Markup Language (XML). The chapter begins with an introduction to UML as a modeling tool very well perceived within the software engineering community. It is followed by a discussion on an attempt by the Object Management Group (OMG) to empower UML for the development of real-time embedded software—an effort which will eventually culminate to a design framework known as the UML Profile for Schedulability, Performance, and Time Specification [29]. Thereafter, an overview of XML and a few other related Internet technologies ensue.

Chapter 4 outlines the structure of the library of platform objects (LPO), as well as furnishes a comprehensive guideline and requirements specification that a platform object must possess in order to be scalable and compatible with the proposed approach. Essential elements for each platform object, e.g., architecture blueprint, XML-based self-described modules, platform managing tool, etc., are also discussed in detail.

Chapter 5 provides a detailed treatment of UML extensions for the development of real-time embedded systems. The chapter starts with a layout of the Co-design Modeling Framework (CMF) hierarchy that encompasses five other sub-profiles—the generic utility profile (PCUprofile) the Exception Modeling profile (EMprofile), the Interrupt Modeling profile (IMprofile), the Synthesizable Hardware Description Language profile (SHDLprofile), and the Architecture Blueprint profile (ABprofile). Each of these profiles furnishes a design framework that is specifically tailored for the proposed approach, and may be able to meet the challenges posed by the design and test of real-time embedded SoC-based systems. The chapter, then, proceeds to discuss the domain concept for each sub-profile, followed by the description of the corresponding stereotypes.

Chapter 6 applies the platform-centric SoC design method, using the CMF profile in UML, to the development of a simplified digital camera system so as to demonstrate the use and the robustness of the proposed approach. Specifically, the NiOS development board is used to mimic the digital camera system where raw image data are read from a charge-coupled device (CCD), and then JPEG encoded and stored into memory. The chapter begins with an overview of the Altera's NiOS system, followed by the actual system development process that explicitly demonstrates the use of the proposed approach. A quantitative evaluation is then presented that compares the development cost of the proposed platform-centric SoC design method against the some alternative approaches using cost estimation models and tools.

Chapter 7 concludes the book with a summary and a discussion of future directions for this effort on platform-centric design.

Chapter 2

PLATFORM-CENTRIC SOC DESIGN METHODOLOGY

2. INTRODUCTION

In the proposed platform-centric SoC method, a library of platform objects (LPO) and the UML work collaboratively to enhance the system development process. While UML manifests itself as a powerful solution to designing and managing complex SoC systems, the scalable LPO aids the developer by elevating the design abstraction level as well as providing a set of pre-characterized constraints and knowledge-based environment for the system developer. This section introduces the platform concept.

2.1 THE PLATFORM CONCEPT

2.1.1 Introduction to Platforms

Karl Sabbagh of Boeing Corporation first defined platforms as fully-functional families of products, each of which is characterized by a set of commonalities, and is specified and implemented in such a way that allows itself and its capabilities to be further customized and re-targeted for specific actual end products. Examples of platforms are abundant, across diversely different application areas, ranging from aircraft to mobile phones. For instance, the Boeing 777 passenger doors, each of which is composed of a different set of parts with subtly different shapes and sizes for its position on the fuselage, are yet based on the same platform, resulting in 98% of all door mechanisms being common [96]. PC platforms, which evolve around the x86 instruction set architecture, a full set of buses, legacy support for the interrupt controller, and the specification of a set of communication devices [14], represent other examples. DARPA's RASSP Program introduced the concept of "model year architectures" in early 1993, which may also be considered equivalent to the current definition of platforms for electronics systems (http://www.eda.org/rassp). The concepts of model year architectures, in turn, were derived from earlier successful models of production of automobiles over a century.

Even though commonality in platforms often could compromise product performance and hinders innovation and creativity, it expedites the overall process of developing end products. A typical platform could spawn scores of products that are more quickly and economically upgradeable through an upgrade of the platform itself. Such advantages are greatly desirable in the development of embedded SoC systems today, where quick time-to-market and ease of upgradeability are the dominating factors [138]. Sangiovanni-Vincentelli and Martin [14] argue that design problems are pushing IC and system companies away from full-custom design methods, towards designs that they can assemble quickly from pre-designed and pre-characterized components. In addition, because platforms can potentially yield high-volume production from a single mask set, manufacturers will tend to be biased towards platform utilization to counter the ever increasing mask and manufacturing setup costs. These platform benefits have become even more attractive as the present state of advances in IC technologies results in more readily acceptable system performances that suit a wide range of applications.

Although the notion of platforms has existed for years, only recently has it drawn a great deal of interest from the electronic systems design community. Currently a number of system platforms exist that can roughly be classified as follows [70]:

- Full-system platform: Platforms in this category often are complete with respect to hardware and software architectures that full applications can be implemented upon, and are generally composed of a processor, a communication infrastructure, and application-specific blocks. Some also utilize FPGAs to attain better flexibility. Examples of full-system platforms include Philips' Nexperia, TI's OMAP multimedia platform (Figure 2.1), Infineon's M-Gold 3G wireless platform, Parthus' Bluetooth platforms, ARM's PrimeXsys wireless platform, Motorola's

Black, Green and Silver Oak, Altera's Excalibur, Quicklogic's QuickMIPS and Xilinx' Virtex-II Pro.

- Quasi-system platform: Platforms in this category generally do not specify full hardware and software architectures so as to provide more flexibility for system developers to re-target them for a wider range of applications. Such platforms as Improv Systems, ARC, Tensilica and Triscend focus more on the ability to configure processors, while others such as Sonics' SiliconBackplane and PalmChip's CoreFrame architectures provide neither a processor nor a full application, but rather define interconnect architectures that full systems can be built upon instead.

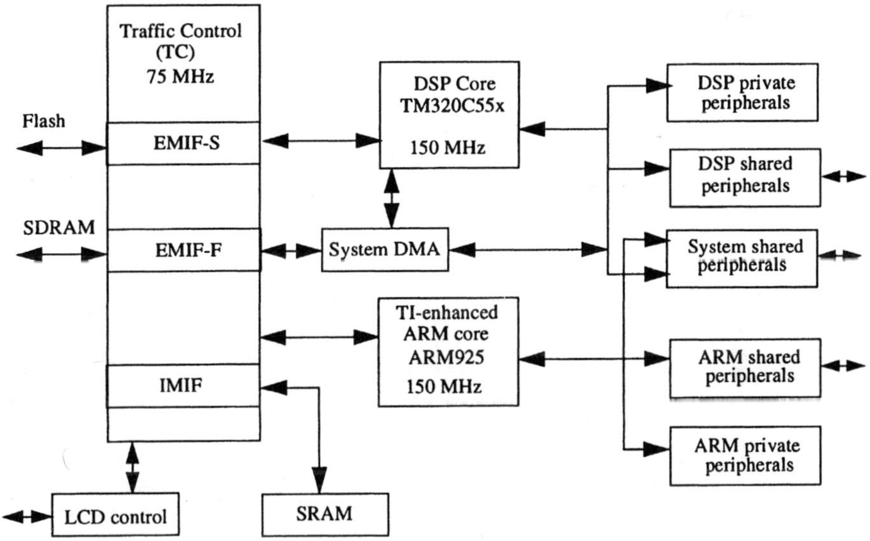

Figure 2-1. A simplified view of TI's OMAP5910 289-pin wireless platform architecture, which has a package size of 12x12 mm2 (based on a figure in [70]).

The system platform should also include the tools that aid the designer in mapping an application onto the platform in order to optimize cost, efficiency, energy consumption, and flexibility.

2.1.2 Platform-based Design for Embedded SoC Systems

The basic idea behind the platform-based design approach is to avoid designing a chip from scratch [138]. The utilization of platforms limits choices, thereby providing faster time-to-market through extensive reuse, but also reducing flexibility and performance compared with a traditional ASIC or full-custom design approach. Goering [70] surveys how the platform-based design is defined, and presents them as follows:

- The definition and use of an architectural family, developed for particular types of application domains, that follows constraints that are imposed to allow very high levels of reuse for hardware and software components (Ref: Grant Martin, Cadence).

- The creation of a stable microprocessor-based architecture that can be rapidly extended, customized for a range of applications and delivered to customers for quick deployment (Ref: Jean-Marc Chateau, ST Microelectronics).

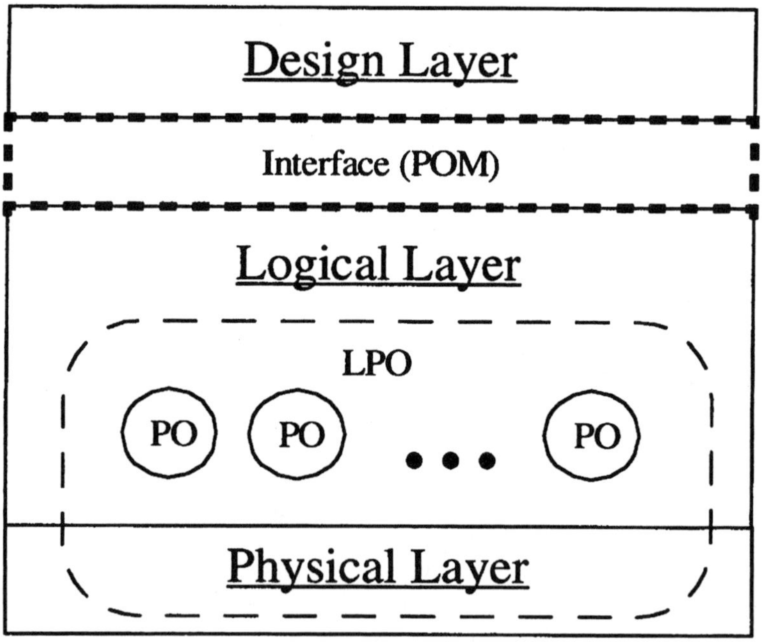

Figure 2-2. A logical model of the platform-centric environment

- An integration-oriented design approach emphasizing systematic reuse, for developing complex products based upon platforms and compatible hardware and software virtual components (VCs), intended to reduce development risks, costs and time-to-market (Ref: Virtual Socket Interface Alliance's (VSIA) platform-based design development working group).

2.2 PLATFORM-CENTRIC SoC DESIGN APPROACH

The platform-centric method is an enhanced version of a platform-based design approach. It provides a tools-integrated environment to assist the designer in coping with the complexity and the various requirements of today's real-time, embedded SoC systems. The approach follows the design flow depicted in Figure 2.3.

Definition 2.1. The Platform-Centric SoC Design Method

A reuse-intensive, software-biased, and analysis-driven co-design approach that relies upon the UML-/XML-enhanced tools-integrated environment and the use of platforms to develop a feasible SoC system quickly and correctly.

Figure 2.2 illustrates the layered architecture of the platform-centric environment. In the physical layer reside the actual hardware and software components, as well as associated design tools, all of which provide the necessary resources for the development of SoC systems.

Definition 2.2. The Platform Object (PO)

For any platform that resides in the physical layer and is a member of a particular library of platform objects (LPO) associated with a platform-centric SoC design method, there always exists a corresponding platform object (PO) in the logical layer that models such a platform.

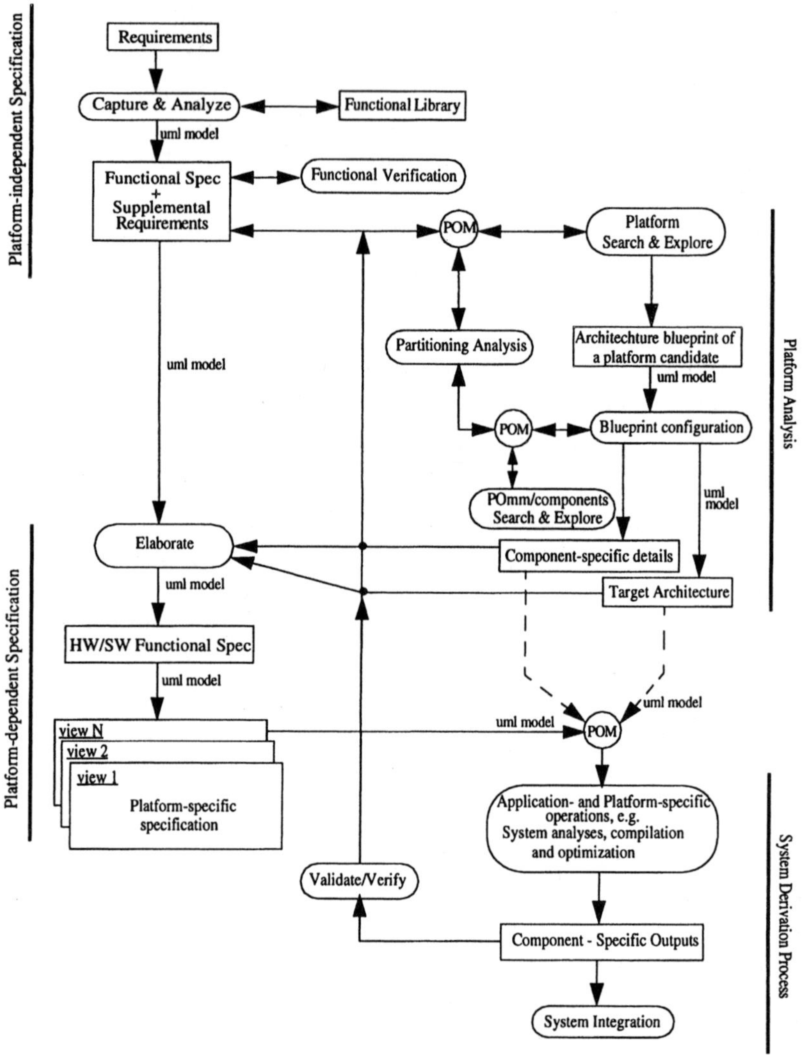

Figure 2-3. The platform-centric SoC method design methodology, utilized by the proposed Co-Design Modeling Framework (CMF)

The logical layer represents a pool of platform models called platform objects (PO), as well as models of other related entities that implement the various platform objects and/or that are platform-compatible and can be used later to build systems based on the proposed approach. Effectively these platform objects and their affiliated modules, such as an abstract representation of the platform, provide an architectural reference for the system developer. Auxiliary information, which uniquely defines and identifies such an abstract representation and the relationship with its counterpart in the physical layer, forms a library of platform objects (LPO). It is the LPO modules in the logical layer with which system developers have direct contact, and not those in the physical layer.

The LPO in the logical layer can be envisaged as a database of the characteristics of platforms and their affiliated modules, and thus, is suitable to be uniformly represented using the Extensible Markup Language (XML). The use of XML also brings forth other benefits, the most important of which arguably is the ability to blend well into the Internet. This capability ensures that the LPO boundary is as expansive as the Internet itself, and these LPO member modules are un-encumbered by geographical location. Furthermore, the LPO can potentially inherit many characteristics of the Internet technologies which can result in each of its PO behaving like a directory that can dynamically grow and shrink with respect to the contents, i.e., the PO member modules (POmm), present at the time of search. The LPO that can change dynamically in such manner is said to be scalable.

System development activities take place in the design layer. The developer accesses the resources in the logical and physical layers through a pre-defined interface called platform object manager (POM). This interface is usually a software tool member of the physical layer.

The platform-centric SoC design approach promotes an enhanced tools-integrated environment while also raises a level of design abstraction. Each platform object (PO) represents a collection of common configurable architectures along with their affiliated components and tools that belong to the platform domain.

A configurable platform is pre-designed off cycle, often optimized for a specific application domain. The platform's common architecture model, the blueprint, fosters the concept of scalability and "parametrizability"; it allows candidate components to be added in or taken out without affecting other candidates. Besides mitigating the architecture selection problem, such a characteristic can help the system developer avoid lower-level hardware-dependent programming, while, at the same time, posing additional requirements for the PO member modules (POmm). As such, a library of platform objects is a logical collection of pre-designed, pre-characterized platforms that are further governed by a set of rules and requirements specific to the proposed approach. Chapter 4 discusses these rules and requirements in detail.

The proposed design flow commences with the requirements capture and processing step that results in the platform-independent functional specification as well as the specification for timing and other requirements. The system developer then applies an estimation technique on the resultant functional specification so as to acquire general performance estimates and uses them to further help select the target architecture. With the target architecture information available, the developer can subsequently derive the detailed platform-dependent specification, which may contain a collection of different analysis models, e.g., concurrency model, subsystem and component model, etc. This specification along with the information relevant to the selected architecture is passed back to appropriate LPO modules to be further analyzed, realized and integrated. Detailed discussion regarding each main stage in the proposed approach is attempted next.

2.2.1 Platform-Independent Specification

This stage chiefly concerns with the derivation of the functional specification that is still independent of any platform instance binding. Unlike most current co-design approaches that begin with a formal specification of the system, the proposed platform-centric method starts with the requirements capturing process. Kotonya and Sommerville [35] define a "requirement" as a statement of a system service or constraint. A service statement describes how the system should behave with regard to the environment; whereas, a constraint statement expresses a restriction on the system's behavior or on the system's development.

The task of requirements capturing typically involves two main subtasks, namely, determining, and analyzing the requirements as imposed by the customer [36]. During the requirements determination subtask, the developer determines, analyzes and negotiates requirements with the customer. It is a concept exploration through, but not limited to, UML's Use Case diagrams. The customer involvement in the requirements capturing process is highly recommended. Once all the requirements are determined, the analysis subtask begins that aims at eliminating contradicting and overlapping requirements, as well as keeping the system conforming to the project budget and schedule.

The functional requirements are then modeled using Use Case diagrams; while, those with non-functional characteristics, e.g. speed, size, reliability, robustness, portability, standards compliance, ease of use, etc., may be captured into the supplementary requirements specification (usually a note or textual description) later to be processed and incorporated into the functional specification as constraints [97].

As established techniques in object-oriented design, the UML's Use Case and Class diagrams play a major role in deriving the platform-independent functional specification. Once all classes are identified, detailed functional implementations ensue. Various UML diagrams may be used to describe the system characteristics.

For example, Sequence and/or Activity diagrams can capture concurrent and/or sequential interactions; or, Component diagrams can depict components connection within the system.

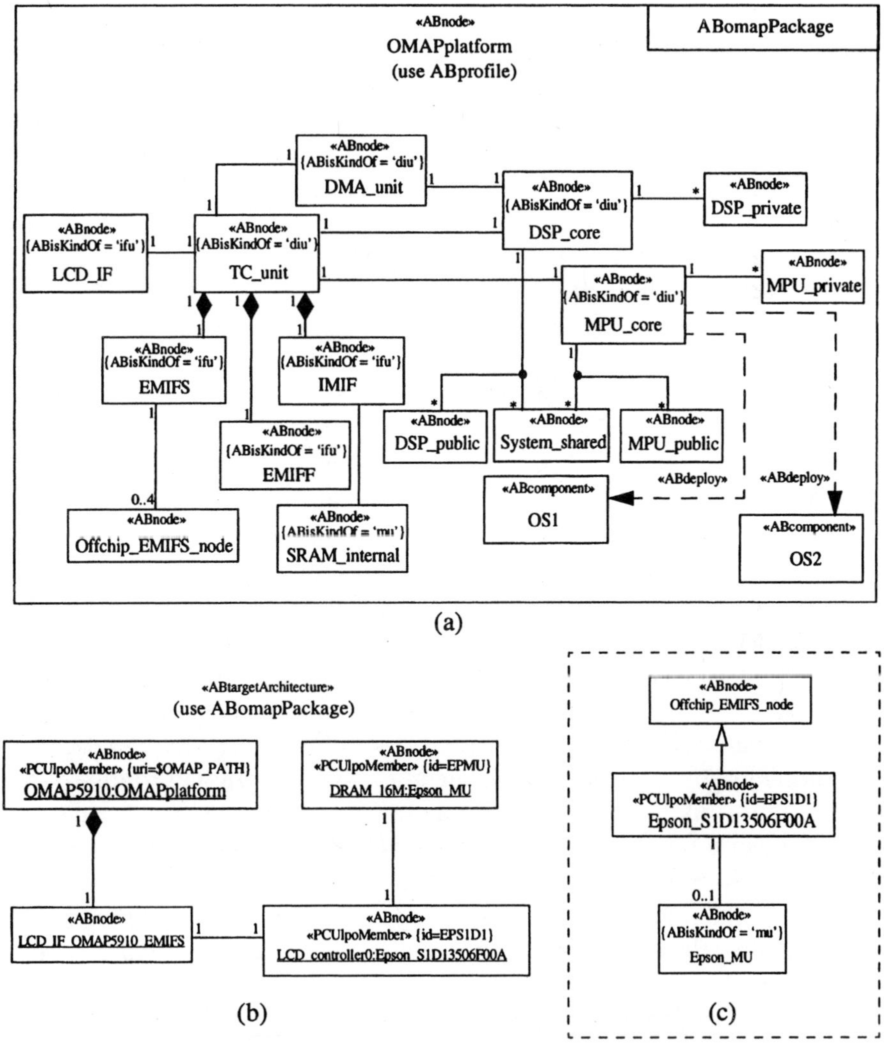

Figure 2-4. TI's OMAP architecture blueprint which (a) depicts the abstract representation of the platform architecture, and is used by Pomm suppliers as a reference model, and by the developer to construct the target architecture (b). Each link in the object diagram (b) represents a pre-defined communication. The DRAM object comes as a derivative requirement when instantiating the LCD controller Pomm module (c)

The resulting functional specification can have the class methods implemented using a programming language of choice. This specification may later be verified for behavioral correctness and handed off to the next stage.

2.2.2 Platform Analysis

The primary objective of this stage is to select the target architecture from the library of platform objects (LPO). Each PO is categorized by its applicable area(s) of use. The developer selects a candidate platform most likely to be successful for the application at hand based on information such as the associated datasheet and the requirements specification obtained in the previous stage. The configuration of the selected platform object then follows, resulting in the target architecture model.

The configuration process starts by acquiring an architecture blueprint specific to the chosen platform object by means of the platform object manager (POM)—a front-end tool equivalent to the interface concept in Figure 2.2. The architecture blueprint is a collection of logical PO member modules (POmm), each of which corresponds to a physical component in the platform. The relationship among the POmm in the blueprint is captured using the UML's Class diagrams—the ABprofile framework to be discussed in Chapter 5. Figure 2.4 (a) illustrates the architecture blueprint of the simplified Texas Instruments' OMAP platform shown in Figure 2.1.

In order to construct the target architecture model, the developer may utilize estimation tools especially optimized for the platform object to provide estimates of design metrics such as execution time and memory requirements for each part of the platform-independent specification. Estimation can be performed either statically by analyzing the specification or dynamically by executing and profiling the design description. The estimation tools take as the input the platform-independent specification, in XML Metadata Interchange (XMI) format which is defined as part of UML, and generate estimation results to be used by the developer and/or by the architecture-selection search algorithm. These results can further be validated against the requirements either manually or using validation tools.

The platform characteristics dictate that a microprocessor or microprocessors be pre-selected and pre-optimized for specific areas of application. This places a considerable constraint on the tasks of architecture selection and system partition, which consequently could result in a faster development time trying to achieve a feasible system. Also such characteristics could permit estimation tools and profiling techniques to perform more accurately, as well as attribute to a smaller search space for the automated algorithms. In the proposed approach, the tasks of selecting the target architecture and partitioning the system are interwoven. The developer can partition the system and/or select the candidate hardware components manually, or by using various architecture selection techniques such as simulated annealing, genetic algorithms, and tabular search. The reader may wish to consult references [23], [91] for detailed treatments of these algorithms on the architectural selection process. In [92], [93], [94], other related algorithms are presented.

It may be noted that, in developing a system where flexibility and technology-to-market time are the predominant factors, software may often carry more weight than the hardware counterpart. A study by Edwards [37] based on Amdahl's Law [38], which limits the possible overall speedup effects obtainable by accelerating a fraction of a program, suggests that it is not worth the effort to attain a substantial speedup for an arbitrary program by moving part of it to hardware. To support his claim, Edwards shows that if a part of the program which accounts for as much as 90% of the execution time is moved to an ASIC, it is still not possible to achieve more than a speedup factor of 10, even under ideal circumstances where the ASIC executes infinitely fast and communication cost is disregarded.

Even though the use of nearly any hardware/software partitioning algorithm is arguably viable for the proposed approach, the developer should always be aware of the increasing weight of software in today's SoC systems development. The platform-centric approach addresses this issue in two ways. Firstly, by utilizing the platform concept as the basis for the design, the proposed approach allows the system developer to concentrate harder on the software development process. Secondly, by utilizing UML as the unified representation of the system under development, the proposed approach allows the task of developing a software system to be handled more robustly by one of the best techniques in the software engineering discipline. In Chapter 5, the UML profile for Co-design Modeling Framework is presented that helps bridge the gap between the platform concept and UML, and effectively, hardware and software.

Once all the hardware components are chosen, they will be (1) instantiated with the blueprint to derive the target architecture, and (2) associated with the hardware-bound software modules, and (3) assigned the proper parameter values either manually by the designer or automatically either by the POM or an accompanying configuration tool. At the end of this stage yields the target architecture which can be used to derive hardware-software functional specification. The target architecture is essentially represented by the concrete instance of the architecture blueprint. Figure 2.4 (b) illustrates the logical representation of the simplified OMAP5910 Video architecture as discussed in [95]. This Video architecture utilizes a LCD controller POmm module, as shown in Figure 2.4 (c), and a 16-MByte DRAM as permissible by the LCD controller.

2.2.3 Platform-Dependent Specification

The XML-based library of platform objects (LPO) is information-rich. Each component in the LPO is self-descriptive; it supplies the developer and the corresponding PO managing tool (POM) with information such as identity, design properties (weight, dimension, times, etc.), tool-related information, and so on. One of the information sets it may carry is the description of hardware-dependent software routines that can be used by the developer to avoid low-level coding of hardware dependent routines. Consider the LCD_controller0:Epson_S1D13506F00A object in Figure 2.4 (b). This object is represented by a specific XML description of the Epson_S1D13505F00A instance—with all the required parameters configured, and pre-characterized data captured. Within this XML description exists a section that describes hardware-dependent software routines, a corresponding device driver and the location where their implementation may be found. The availability of these functions facilitates the software development process by helping the system developer avoid implementing specific hardware-interface routines—it virtually hides communication-related details at a lower level of design abstraction.

The platform-dependent specification stage is iterative, and during this phase, the developer first derives the hardware-software functional specification. Thereafter, additional models, e.g. concurrency model, schedulability model, etc., may be developed and analyzed (in the next stage), whose results are back-annotated for further elaboration of the hardware-software functional specification. Such activities may proceed iteratively, as shown in Figure 2.3, until all required system characteristics are determined, at which point the implementable platform-dependent specification results that can now be realized and integrated into a full system prototype.

2.2.4 System Derivation Process

The System Derivation process is carried out iteratively, hand in hand, with the other stages; it furnishes an execution domain for the platform-dependent models resulted from the preceding Platform-Dependent stage. The tools and tasks involved within this stage vary considerably ranging from detailed analyses, to validations and verifications, to hardware and software syntheses, to system integration. Software modules may be compiled, optimized and saved in a microprocessor-downloadable format, and hardware-bound modules may be synthesized and saved into some common format like the EDIF netlist file. Or, a schedulability analysis tool may take a partially specified schedulability model and completes it for the developer—a process known as parameters extraction. At the end of this stage, accurate timing and other constraints information can be obtained and verified/validated that leads to the system integration.

At this stage, the platform object manager (POM) simply functions as an auxiliary tool that supplies the system developer with necessary information to aid a smooth flow of the development process. In its simplest form, it behaves just like a web browser with all relevant information compiled and readily retrievable. A more sophisticated POM may permit other tools to be called and operated within itself, rendering a highly integrated environment of tools and information.

2.3 COMPARISON WITH CURRENT APPROACHES

Table 2.1 shows how the proposed approach compares with previous related approaches in current practice, albeit in the research domain. The general framework for the hardware/software co-synthesis approach often involves a pre-determined hardware architecture consisting of a particular microprocessor and an ASIC, and the hypothesis that if only the behavior could be partitioned between these components, the remainder of the design process would automatically be done by the high-level synthesis tool and the compiler [23]. A representation of the co-synthesis approach includes the work by Ernst, et.al. [39]. They describe the COSYMA (Co-synthesis of eMbedded Architectures) system which is used to extract, from a given program, a slice of functionality that could be implemented in an ASIC. COSYMA uses a novel software-oriented HW/SW partitioning algorithm that identifies critical operations in an instruction stream and moves those operations from software to hardware. Gupta and De Micheli [40] work with a similar architecture, but take the inverse approach. Instead of trying to accelerate a software implementation, their Vulcan co-synthesis system aims at reducing the cost of an ASIC design by moving less time-critical parts to a processor—checking timing constraints and synchronism as it does so. Yen and Wolf [41, 42] also propose efficient algorithms for co-synthesizing an embedded system's hardware engine, consisting of a heterogeneous distributed processor architecture, and application software architecture, consisting of the assignment and scheduling of tasks and communication of an embedded system. Although these excellent co-synthesis approaches represent state-of-the-art efforts in co-design, they do have certain limitations—some of the more notable ones are:

- In the COSYMA and Vulcan systems, the fixed architecture is presumed without elaborating on such a decision. This could the scope of applicability of these approaches.

- Yen and Wolf's approach supports only bus-oriented architectural topologies, which may not scale well for large application.

- Automated architectural selection is only partially supported in Yen and Wolf's approach, and not at all in COSYMA and Vulcan systems. They are unable to raise design-abstraction level and/or capable of capturing customer's requirements. Neither is there a support for real-time handling mechanism.

The model-based approach [43, 44] fosters a late-partitioning, late-technology binding philosophy. Its basic supposition is that models serve as design blueprints for designing systems. At its core, the model-based approach employs a modeling technique to capture systems behaviors at different levels of abstraction. The resulting models are then subject to validation and stepwise refinement process. A validated model is simulated in a specific set of experimental conditions to verify its adherence to the initial requirements, constraints, and design objectives. Technology assignment is then carried out from the verified model specifications. The approach handles the complexity issue via the use of design modularity and hierarchy where the designer constructs models from elementary building blocks that are connected into larger blocks in a hierarchical manner. Nevertheless, the approach has some limitations:

- The approach does not explicitly specify the synthesis process.

- Lack of support for translation process and tools makes it insufficient to handle the requirements of systems today.

- There is no real-time handling mechanism, nor the capability to capture customer requirements.

The POLIS [11] co-design method addresses the issues of unbiased specification and efficient automated synthesis through the use of a unified framework, with a unified hardware-software representation. POLIS is an example of the tools-integrated environment that relies on performance estimates to drive the design, and on automation techniques to complete the tasks at each design step. The integral idea behind POLIS is the Co-design Finite State Machine (CFSM).

Table 2-1. Feature support of current co-design approach: "supported" (++), "partially supported" (+) and "not supported" (BLANK). The survey approaches include the Model-based [43, 44], POLIS [11], Corsair [10], SpecC [20], SystemC [131], Chip-in-a-day [46]

Features	Model-Based	POLIS	Corsair	SpecC & SystemC	Chip-in-a-day	Platform-Centric
Mechanism for customer's requirements determination and analysis						++
Mechanism for constraints capturing and validation	+	+	++	+		++
Mechanism for real-time handling		+	++	++		++
Unified representation	++	++	++	++	++	++
Elevation of design abstraction level	++	+	+	+	+	+
Translation and elaboration	+	++	++	++	++	++
Tools integration environment		++	++	++	++	++
Feasibility for large-scale systems design	++	++	++	++		++
Suitability for wide range of application	+	++		++		++
Being adaptive to technology-specific approaches						++
Integrated documentation						++

The CFSM provides a unified input for the tools within the POLIS environment. POLIS supports automated synthesis and performance estimation of heterogeneous design through the use of Ptolemy [45] as the simulation engine. Such ability allows POLIS to provide necessary feedback to the designer at all design steps. A simple scheme for automatic HW/SW interface synthesis is also supported.

The Corsair [10] integrated framework methodology is similar to POLIS. The Corsair framework contains several tools for the automated implementation of formally specified embedded systems. Based on the extended specification language Specification and Description Language with Message Sequence Chart (SDL/MSC), the approach supports system synthesis, implementation synthesis and performance evaluation for rapid prototyping. The data from performance evaluation are back-annotated to support the estimation tool during the system synthesis step.

Even though POLIS and Corsair represent a general improvement for the design of complex embedded SoC systems, several limitations still exist:

- The CFSM graph used in the POLIS framework provides very little support for real-time handling mechanism.

- POLIS's automatic interface generation scheme is still very primitive. Much work has to be done for it to be able to cope with large-scale systems design.

- Although the tools in the Corsair environment are very well integrated, it only works for a fixed architecture. The approach gives no insight why this particular architecture is selected.

- Both methods do not furnish a process for capturing customer's requirements.

The SpecC method [20], on the other hand, is based on a specify-explore-refine paradigm. It is a unified language, IP-centric approach aimed at easing the problems caused by heterogeneous design. The SpecC language can be employed to describe both hardware and software behaviors until the designer attains the feasible implementation model later on in the design process, hence, promoting an unbiased hardware/software partitioning for the system under development. The formalism of the SpecC language allows for efficient synthesis. SpecC provides support for exceptions handling mechanism, and is capable of capturing information about timing constraints explicitly within its constructs. Limitations of the SpecC method are in the following areas:

- SpecC only supports the capturing and propagation of timing constraints, but not other constraints. This makes the tasks of constraints validation more difficult.

- The approach does not specify a process and a mechanism to systematically handle customer's requirements.

Similar in motivation to SpecC, a new language called SystemC has been proposed [131]. Based entirely on C++, SystemC provides a unified, IP-centric environment for specifying and designing SoC systems, and is capable of modeling systems at different abstraction levels from the transaction-accurate level to the bus-cycle accurate level to the RTL level for hardware implementation—making it well-suited for an interface-based approach that fosters co-simulation and/or co-verification of heterogeneous systems under development. The SystemC's support for executable specification modeling within a single language facilitates HW/SW co-design while, simultaneously, easing the co-simulation/co-verification tasks. In addition, its generalized communication model permits an iterative refinement of communication through the concepts of ports, channels, and interfaces. Other features of SystemC include fixed-point modeling capability, and easy integration of existing C/C++ models.

Nonetheless, SystemC does have its own limitations that manifest in the following areas:

- Like the SpecC approach, SystemC does not specify a process and a mechanism to systematically handle customer requirements.

- SystemC is relatively new, especially as a hardware description language (HDL). The current version (SystemC 2.0) has yet to provide support for hardware synthesis. The language lacks also the wide industry and engineering backing that the UML enjoys.

- The interface-based approach adopted by SystemC singularly might not suffice to tackle current issues in the development of SoC systems.

The chip-in-a-day approach [46] proposed by the University of California, Berkeley's Wireless Research Center (BWRC) represents an early prototype for the more promising platform-based design [14]. The approach uses Mathworks' Simulink to capture a high-level data flow and control flow diagrams. Based on pre-characterized hardware components, it implements data path macros directly using a tool such as Synopsys's Module Compiler, while the control logic is translated into VHDL and synthesized. In this approach, algorithms are mapped directly into hardware that derives its parallelism not from multiple CPUs, but from a multitude of distributed arithmetic units. Although the current results claim to be two to three orders of magnitude more efficient in power and area than architectures based on software processors, the chip-in-a-day approach finds its limitations in the following:

- It still cannot improve upon the execution time.

- It remains to be seen if this approach can support a much more complex real-time embedded SoC design.

2.4 OTHER EMBEDDED DESIGN APPROACHES USING UML

The platform-based design concept [14] upon which the chip-in-a-day approach is based also spawns an inception of the proposal for the Embedded UML profile [133] whose objective is to merge real-time UML and HW/SW co-design together. Embedded UML coalesces various existing ideas currently used in real-time UML and HW/SW co-design practices. It is conceived by its initiators as a UML profile package which is "suitable for embedded real-time system specification, design, and verification" [133].

In a nutshell, Embedded UML models the system using a collection of reusable communicating blocks similar to ROOM's capsule concept [134]. Interfaces and channels, that are the extensions of ROOM's port and connector notations, are used for communication specification and refinement. Within these interfaces and channels, UML stereotypes can be defined for communication and synchronization services. Other UML models such as Use Case and Sequence diagrams provide means for specifying test benches and test scenarios, while a combination of well-defined State diagram semantics and Action semantics [133] serves as a driving force for code generation, optimization, and synthesis. In addition, the extended Deployment diagrams, called the Mapping diagrams, may be employed to model system platforms [14] and furnish the platform-dependent refinement paradigm for performance analysis, and optimized implementation generation.

Even though no detailed implementation of the Embedded UML is available at the time of this writing, a careful perusal over its proposal reveals a few interesting facts. The Embedded UML profile and the UML profile for Co-design Modeling Framework (see Chapter 5) bear some resemblance as per their underlying objectives—each of which attempts to furnish a means to model and to develop platform-based real-time embedded systems. Certain minute implementation details of the two profiles differ tentatively. While the Embedded UML resorts to the ROOM's real-time modeling approach [134] and utilizes capsule-based reusable blocks as the basic design units, the Co-design Modeling Framework profile relies on the UML profile for Schedulability, Performance, and Time Specification [29] and perceives all design entities, including communication links and protocols, as reusable objects in the LPO. Yet another fundamental difference between the two profiles exists that possibly comes as a consequence of discrepancy in the viewpoint towards how each of them should be conceived. Just like Vissers [9], many propose that semiconductor systems manufacturers & integrators will design systems, as opposed to designing processors for the short term, though it may be to their advantage in the longer term to design their own processors. The proposed UML profile for Co-design Modeling Framework is flexible to meet with new demands from these organizations as they adapt to new technologies and business models. This is to contrast to the Embedded UML profile whose approach seems to come from the opposite direction where system houses are seen as the primary forces in the development process. Under such a circumstances, it is likely that new methods and new ways defined by the semiconductor sector to build systems could possibly prevent the Embedded UML from attaining its full effectiveness—mainly due to the generalization of the Embedded UML package. As a preliminary assessment, the approach embraced by this book may eventually be more robust in a long run as it is designed to better adapt to technological changes.

Proposed as part of the Yamacraw Embedded Software (YES) effort at the Georgia Institute of Technology's Georgia Electronic Design Center (GEDC), the YES-UML [135] represents yet another UML extensions package that aims at furnishing the system-level unified representation for the development of today's embedded systems. The YES approach fosters the concept of extreme reuse, where the development process encompasses both the Application Engineering and the Domain Engineering arenas, and the efficiency gain results from the application of the economies of scope [136] that suggest the development of a family of products rather than a single product as traditionally practiced.

The YES-UML is perceived as a complete system integrating notations whose objective is to "address the multiple-language, multiple-analysis problem of embedded systems design by combining together levels of abstraction and heterogeneous conceptual models. [137]". Owing to its unified representation capability, systems analysts can deal directly with the UML models at the front-end, whereas systems designers can utilize their conventional analysis tools seamlessly at the back-end through the XMI interface. The proposed YES-UML extensions encompass the following modeling capabilities for [135]:

- Behavior of analog interfaces within the embedded specification,
- Real-time related behavior within the specification,
- Early identification of size, weight, and power (SWAP) constraints,
- Hardware/software interfaces, and
- The development of an executable specification capable of expressing concurrency in the real-time system being described.

It may be clear from this discussion that, like the Embedded UML, the YES-UML also bears a number of similarities to the Co-design Modeling Framework (CMF) profile even though its intended development paradigm is larger in scope than that of the proposed platform-centric methodology.

2.5 A PERSPECTIVE ON COLLABORATION WITH NON-PLATFORM APPROACHES

Being a graphical language, the UML works well for the proposed approach to help promote reuse at a high abstraction level—specifically at the platform level where the system could be quickly assembled using pre-designed, pre-characterized platform components. The proposed approach normally benefits from such platform-component reuse, as well as the use/reuse of knowledge brought forth by the XML technologies, to expedite the overall SoC systems development process. Nevertheless, the diverse requirements of today's SoC systems make it nearly impossible that a desirable platform component would always exist for the system developer when applying the proposed platform-centric SoC design method. As a result, this section discusses a possible collaboration of the platform-centric approach and other non-platform approaches that could spawn an efficient sub-process within the proposed approach, and satisfactorily address this very issue.

A programming language, such as SpecC or SystemC, is particularly well-suited for implementing the functionality of the UML models. The reasons are that (1) both of them are object-oriented, as is UML, which could result in a convenient mapping scheme between the model and the implementation, and vice versa, and (2) like UML, they can uniformly represent hardware and software in a single language. SystemC, in particular, permits a large repertoire of C/C++ reusable modules to be incorporated should need arise. As a result, SystemC manifests itself as a possible language of choice for the proposed approach for the implementation of the UML models.

Figure 2.5 illustrates possible synergy between the proposed UML-based approach and the SystemC approach. In this figure, SystemC is used to implement the functionality of the platform-independent specification resulted from the requirements analysis. In a typical platform-centric development process flow, further analysis on this specification (in the platform analysis phase) will help determine the target architecture as well as the partition of hardware and software modules. At this point, if there exists no desirable hardware component such that the developer has to implement it as part of the design cycle, s/he can conveniently apply the

SystemC iterative refinement approach to the platform-independent functional model—yielding either the SystemC HW/SW model or the RTL netlist which can be inserted back into the platform-centric environment. Such a collaborative approach presented herein provides a safeguard for the proposed platform-centric SoC design method where it can still benefit from a unified design environment and extensive reuse at the interface level, even if/when the platform-level design is not possible.

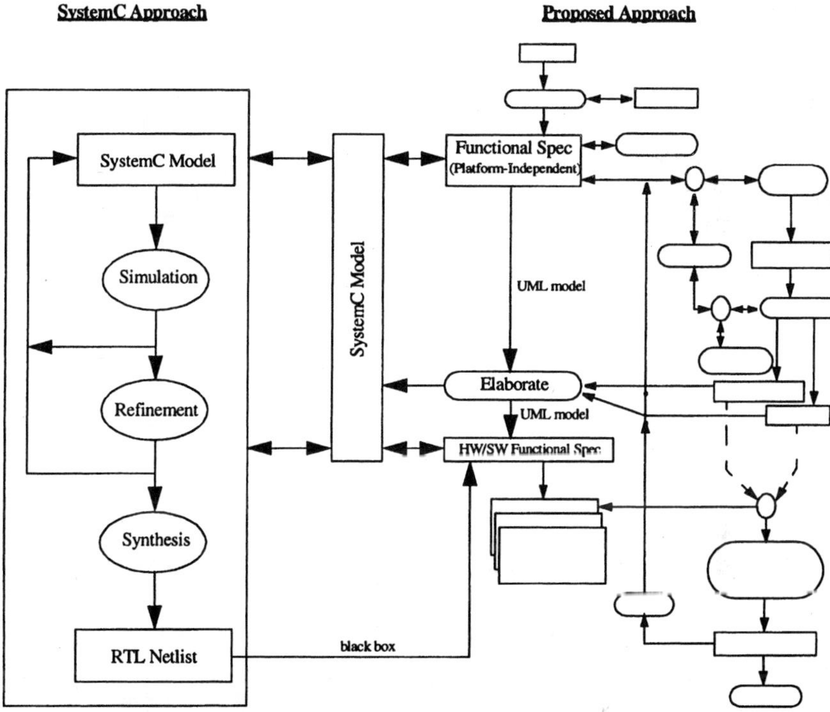

Figure 2-5. Collaborative usage model for the proposed platform-centric approach (CMF) and the SystemC approach (adapted from [132]).

One of the major advantages of the proposed platform-centric SoC design methodology appears to be the creation of an information-rich environment that promotes easy and effective architecture selection process, while at the same time allows a wide range of state-of-the-art tools and technique to co-exist collaboratively, with minimal modification required. In addition, the proposed approach defines a UML framework for the design of real-time embedded SoC systems that begins with customer's requirements determination and analysis, and results in a detailed functional specification. This book asserts that such contributions bring about a desirable paradigm shift along with an improvement in the overall efficiency of the proposed SoC design approach.

The next chapter gives a technological overview of the essential background related to the Unified Modeling Language (UML) and the Extensible Markup Language (XML). It commences with an introduction to UML, followed by a discussion on the UML Profile for Schedulability, Performance, and Time Specification [29]. Thereafter, an overview of XML and a few other related Internet technologies applicable to SoC design is given.

Chapter 3

INTRODUCTION TO UML AND XML

3. INTRODUCTION

This chapter presents an introduction to the Unified Modeling Language (UML) and the Extensible Markup Language (XML) technologies that provide the foundation of the proposed Platform-Centric SoC design methodology. To begin, the chapter explains the UML and its various features, as traditionally used within the software engineering community where UML is originated. It also discusses UML usage scenarios, extensibility mechanisms and possible mappings of UML notations to other programming languages such as Java, and C/C++. Thereafter, it gives an overview of the UML Profile for Schedulability, Performance, and Time Specification. The chapter concludes with a discussion of XML and related technologies that are essential for the robust implementation of a scalable and Internet-oriented library of platform objects (LPO) for the proposed methodology.

3.1 UNIFIED MODELING LANGUAGE (UML)

In the proposed approach, the UML not only serves to drive the development process from its earliest stage of requirement capturing and analysis to the later stage of acquiring the correct implementation models, but it also provides a common input format, i.e. XMI, for the collection of tools in the library of platform objects (LPO). The UML extension mechanism makes it flexible to be adapted to better suit new tools and techniques. In addition, UML can be used for documenting the design.

The Unified Modeling Language (UML) is a language for specifying, visualizing, constructing and documenting the artifacts of both software and non-software systems. It represents a collection of the best engineering practices that have proven successfully in the modeling of large and complex systems [24]. UML bases most of its foundation on the works of Booch, Rumbaugh, and Jacobson, but its reach has come to encompass a greater expanse than originally perceived by its creators [25].

The language has successfully undergone the standardization process with the Object Management Group (OMG) and become widely received by the industry.

The UML is a simple and generic notation made of model elements that can be used to model requirements for design of the system. Mathew [26] describes a UML model as the basic unit of development which is highly self-consistent and loosely coupled with other models by navigation channels. A model is not directly visible to users. It captures the underlying semantics of a problem, and contains data accessed by tools to facilitate information exchange, code generation, navigation, etc. UML models are represented graphically. Many different perspectives can be constructed for a model—each shows all or part of the model and is portrayed by one or more diagrams. Table 3.1 lists the models and diagrams as defined by UML.

Use case diagrams describe what a system does from the standpoint of an external observer. They are closely connected to scenarios—an example of what happens when someone or something interacts with the system. A use case is a summary of scenarios for a single task of goal. An actor is who or what initiates the events involved in that task. The connection between actor and use case is called a communication association. A use case diagram can also be viewed as a collection of actors, use cases and their communications. Use case diagrams are helpful in such tasks as communicating with clients, and capturing and determining requirements.

Table 3-1. UML models and diagrams

UMLModels	UML Diagrams
1. Class model	1. Class diagrams
2. State model	2. Sequence diagrams
3. Use case model	3. Collaboration diagrams
4. Interaction model	4. Object diagrams
5. Implementation model	5. Statechart diagrams
6. Deployment model	6. Activity diagrams
	7. Use case diagrams
	8. Component diagrams
	9. Deployment diagrams

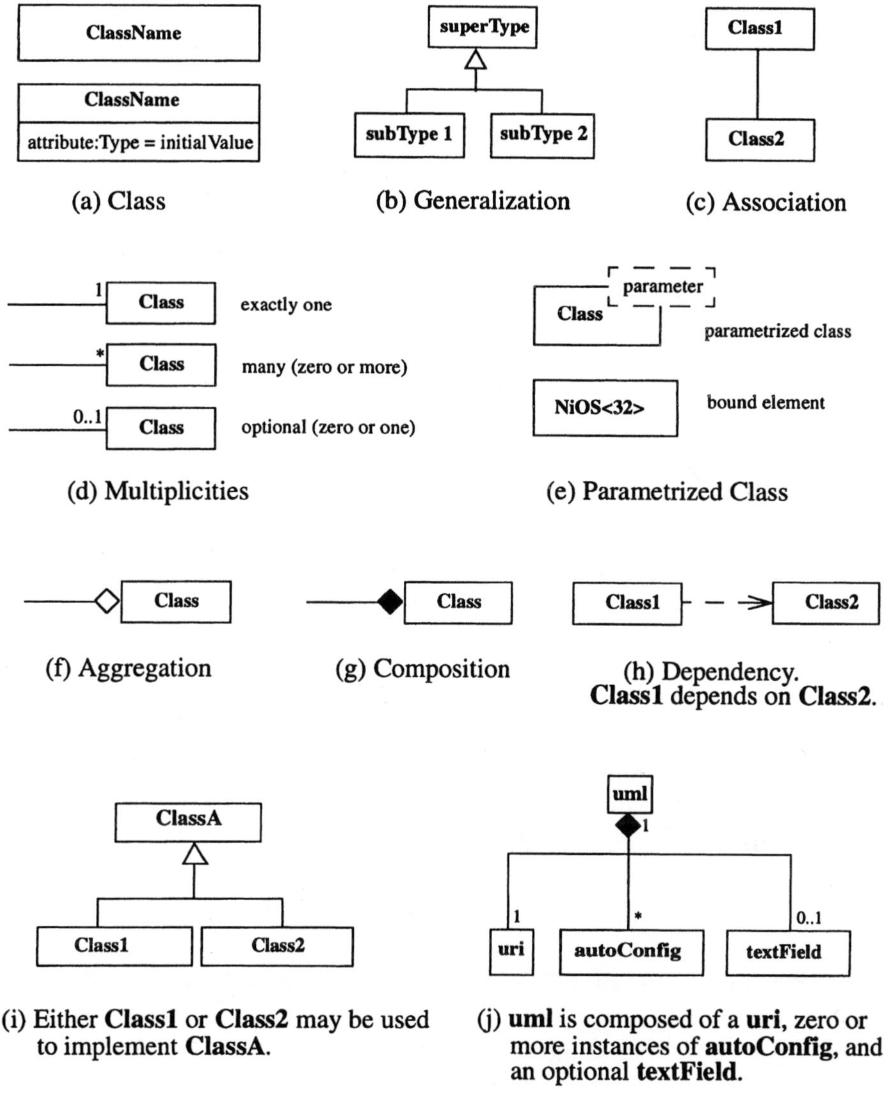

Figure 3-1. Summary of UML notations

A class diagram gives an overview of a system by depicting classes and the relationships among them. Class diagrams are static, which means they display what interacts but not what happens when they do interact. Some useful relationships are, for example, association, aggregation, composition and generalization. A number of attributes can embellish an association: a role name to clarify the nature of the association; a navigability to show which direction the association can be traversed; a multiplicity to govern the number of possible instances of the class associated with a single instance on the other end, and so on. Mandatory elements in each class diagram consist of classes, associations, and multiplicities. Figure 3.1 illustrates a subset of UML notations that are often used in this book and elsewhere.

To simplify complex class diagrams, classes can be grouped into packages. A package is a collection of logically related UML elements. A package is said to depend on another if changes in the other could possibly force changes in the first. Dependency is denoted by a dotted arrow, with the arrow originates from a client package, i.e. a package which depends upon the other package (the supplier) where the arrow terminates. Another UML mechanism that assists in simplifying complex class diagrams is the object diagram. Object diagrams show instances instead of classes, and are useful for explaining small pieces with complicated relationships, especially recursive relationships.

While class and object diagrams are static model views; sequence, activity, statechart, and collaboration diagrams represent dynamic characteristics. A sequence diagram details how operations are carried out, what messages are sent, and when. Sequence diagrams are organized according to time. The time progresses vertically down the page. The objects involved in the operation are listed from left to right according to when they partake in the message sequence. A message in a sequence diagram is asynchronous if it allows its sender to send additional messages while the original is being processed. Similar to sequence diagrams, collaboration diagrams convey the same information but instead focusing on object roles rather than times that messages are sent. In collaboration diagrams, object roles are the vertices and messages are the connecting links with proper sequence numbers on them.

Objects have behaviors and a state. The state of an object depends on its current activity or condition. A statechart diagram shows the possible states of the object and the transitions that cause a change in state. States in statechart diagrams can be nested. Related states can be grouped together into a single composite state. Nesting states is necessary when an activity involves concurrent or asynchronous sub-activities. While a statechart diagram focuses attention on an object undergoing a process, an activity diagram focuses on the flow of activities in a single process. Out of each activity comes a single transition, connecting that activity to the next activity. A transition may branch into two or more exclusive transitions, or merge to mark the end of the branch. A transition may also fork into two or more parallel activities. These parallel threads can later join to form a single transition.

The last two diagrams defined by UML are component and deployment diagrams. Component diagrams are physical analogs of class diagram.

Deployment diagrams display the physical configurations of software and hardware. The physical hardware in a deployment diagram is made up of nodes, which, in turn, can contain components.

UML is more complete than other languages in its support for modeling complex systems, and is particularly suited for capturing real-time embedded systems. The major features of UML suitable for modeling real-time embedded systems include [28]:

- an object model (incorporating data attributes, state, behavior, identity and responsibility) allowing the structure of the system to be captured,

- use case scenarios—allowing key outputs to be identified from system in response to user input,

- behavioral modeling—use of statechart diagrams helps facilitate the dynamic modeling of the system's behavior,

- packaging—providing mechanisms to organize elements of the modeling into groups,

- representations for concurrency, communication and synchronization for modeling real-world entities,

- model of physical topology—using deployment diagrams to show the devices and processors which comprise the system,

- support for object-oriented patterns and frameworks—allowing common solutions to common problems to be represented.

In the context of object oriented technology, patterns can be thought of a problem-solution pair. They capture the static and dynamic structures and collaborations among key participants of successful solutions to problems that arise when building applications in a particular domain [67]. As opposed to frameworks that support reuse of detailed design and code, patterns facilitate reuse of successful software architectures and designs by relying on two key principles [67]: (1) separation of interface from implementation, and (2) substitution of variable implementations with common interfaces. Gamma, *et.al.*, show in their pioneering work on design patterns [68] how patterns solve design problems and how they explicitly capture expert knowledge and design tradeoffs, and make this expertise more widely available.

For further readings on UML, UML's specification by OMG [24] and the user guide by Booch, Rumbaugh and Jacobson [30] provide a detailed treatment of the subject. For patterns, the Design Patterns book by Gamma, et.al. [68] is a good starting point for interested readers. Quick tutorials can be found at [66, 67], while [64, 65] represents the very foundation of design patterns as conceived by Christopher Alexander during the late 1970s [66].

The following subsections discuss UML's general extensibility mechanisms that include topics on constraints and Object Constraint Language (OCL), tagged values, and stereotypes. Then, a discussion on UML-to-code translation, as well as an overview of the UML profile for Schedulability, Performance and Time Specification [29] are presented. This section owes a large part of the UML diagrams description to a good tutorial by TogetherSoft [27].

3.1.1 Constraints and Object Constraint Language (OCL)

To quote [98]: "A constraint is some additional restriction (above the usual UML well-formedness rules) applied against a modeling element." UML constraints always appear enclosed in a pair of curly braces ({ }) and can be placed inside text notes. Certain kinds of constraints, e.g. {subset}, {ordered}, {xor}, etc., are predefined, whereas others can be user-defined (see UML Specification [24] for details). The interpretation of user-defined constraints are tool-dependent. In fact, as stated in the specification, it is expected that individual tools would provide one or more languages in which formal constraints could be written [24].

As the predefined UML constraints are not comprehensive enough to handle all aspects of a system specification, and user-defined constraints can potentially result in ambiguities, UML decidedly includes with its specification a formal constraint description language—the Object Constraint Language (OCL) [24, 99]—that can come in handy when need arises.

OCL expressions are typically used to specify invariant conditions, as well as pre- and post-conditions that must hold for the system being modeled. They can also be used to describe guard conditions, to specify constraints on operations, as well as to provide a navigation language. OCL expressions can reside in text notes, or can be hidden by tools. Because an evaluation of an OCL expression is instantaneous, it never alters the state of the system. Figure 3.2 demonstrates the use of UML constraints and OCL expressions.

3.1.2 Tagged Values

To avoid inundating UML models with an excessive number of graphical notations that often result in developmental ineffectiveness, detailed UML element properties may be captured using such mechanisms as attributes, associations and tagged values.

A tagged value is a keyword-value pair of type String (in the UML specification, a keyword is actually called a tag) that permits arbitrary information to be attached to any kind of model element so as to provide semantic guidance for back-end tools such as code generators, and report writers [24], to name a few. Each tag represents a particular kind of property applicable to one or many kinds of model elements. Similar in nature to UML constraints, a tagged value also always appears within a pair of curly braces ({ }). When more than one tagged value are to be specified, a comma (,) is used as a delimiter to separate them. When a tag appears without an accompanying value, it represents a standard shorthand notation for {isTagname = true}. As an example in Figure 3.2 above, {abstract} is a tagged value that is semantically identical to {isAbstract = true}.

3.1.3 Stereotypes

Among the core extensibility mechanisms furnished by UML, stereotypes are probably the most powerful and extensively used construct. When applied to a model element, it subclasses that model element, inheriting the attributes and relationships, but exhibiting specific intent such that an unambiguous interpretation can be rendered when processed by tools. Just like any model element, a stereotype can have constraints and tagged values attached to itself.

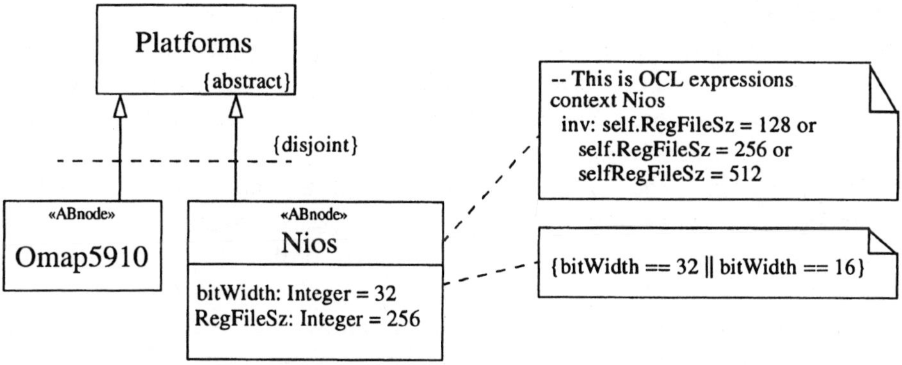

Figure 3-2. Demonstrative use of UML extensibility mechanisms

Stereotypes are usually shown as text imbedded in a pair of guillemets. In Figure 3.2, for example, «ABnode» is a stereotype. So are «ABcomponent», «ABlpoMember», «ABmap» and «ABdeploy» in Figure 2.4. Where the use of guillemets is not possible, the text strings << and >> are used.

3.1.4 UML to Code Mapping

Even though the UML's lack of rigorous formalism has been a subject of criticism by many, it has also been demonstrated by quite a few UML tool vendors [100, 101, 102, 103, 104, 105] that such a shortcoming does not really hinder the developments of features like code generation, reverse engineering (where a tool constructs UML models based on existing code), as well as round-trip engineering, which further tries to regenerate the source code when the model is modified.

Following simple rules like those shown in Table 3.2, the mappings between UML and a popular OO programming language such as Java have evolved to become a standard feature supported by almost all commercial UML tools, plus a couple of free ones. With some modifications, UML to code mappings for other programming languages, e.g. Ada, Real-Time Java, Delphi, CORBA IDL, optimized C/C++ for embedded application, are also possible.

3.1.5 UML Profile for Schedulability, Performance and Time Specification

Currently undergoing the final stage prior to being standardized by OMG, this real-time profile aims to bridge a gap between the real-time and UML communities by providing capabilities for modeling real-time systems and characteristics, such as quality of services (QoS) that often are non-functional, yet essential to the development of real-time systems. The profile furnishes a design framework that (1) enables the construction of models that could be used to make quantitative predictions about the QoS characteristics, and that (2) facilitates communication of design intent among developers in a standard way [29].

A UML profile, shown as a package notation adorned with «profile», is a collection of predefined stereotypes that is used to tailor UML for a specific problem domain. As such the UML profile for Schedulability, Performance and Time Specification (or, contractively writing, the UML real-time profile) provides the system developer, as well as tool vendors, e.g. an analysis method provider, or an infrastructure provider, with a means to effectively model real-time applications for analysis and synthesis via a predefined set of RT-related stereotypes.

At the core of this profile is the General Resource Modeling framework which encompasses the RTresourceModeling, RTtimeModeling, and RTconcurrencyModeling profiles that define the concept and the modeling of resources, time and concurrency, respectively. Along with these three packages, the UML real-time profile also includes the basic frameworks for schedulability and performance analyses, i.e. the SAprofile and the PAprofile packages, as well as the Real-Time CORBA infrastructure model. Figure 3.3, which is adapted from [29] and [106], illustrates the structure of the UML real-time profile that also depicts the interactions among different active entities (represented by "stick figures") and profile packages.

The UML real-time profile's general resource model deals specifically with various aspects of resources modeling in the real-time domain. It details a comprehensive treatment on resources-sharing where mechanisms for modeling activeness and protection are defined. Services as provided by the resources are either exclusive or non-exclusive, with various quality of service (QoS) characteristics imposed upon them.

The time model described in the RTtimeModeling profile is a specialized resource model, and thus, naturally inherits all the rudimentary concept defined in the RTresourceModeling package. Using the RTtimeModeling profile, the developer can specify time value, instant, and duration, as well as associate time with actions and events where moment of occurrence and duration are of importance. It is to be noted, nonetheless, that the RTtimeModeling profile only permits one clock to be defined.

The last package in the general resource framework, the RTconcurrencyModeling profile, deals primarily with the concurrency aspects of modeling real-time applications. Using this profile, the concept of active objects and threads can be modeled conveniently. The profile utilizes a message passing scheme for communications.

Table 3-2. General rules for the mappings between UML models and Java.

UML constructs	JAVA constructs
1. Attribute	1. Instance variable
2. Operation	2. Method
3. Abstract class	3. Abstract class
4. Interface	4. interface keyword
5. Package	5. Package declaration
6. Subclass (Generalization)	6. extend keyword
7. Realization	7. implement keyword
8. «use» or Dependency	8. import clause
9. Multiplicity	9. Array
10. Role	10. Instance variable of type Class associated with the Role

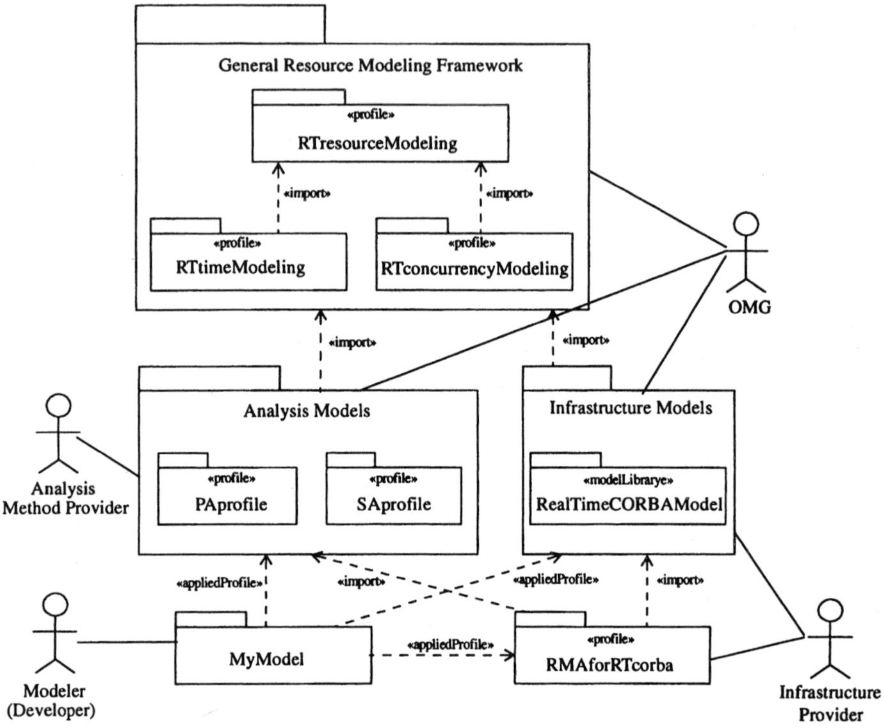

Figure 3-3. Structure of the UML Profile for Schedulability, Performance, and Time Specification

From the profile viewpoint, a call action in the sender initiates a message transfer that subsequently triggers a behavior in the receiver. The modeler can specify call actions to be synchronous or asynchronous, and operations or receptions of messages to be immediate or deferred.

In addition to the profiles discussed earlier, the UML real-time profile also contains two other native packages that provide basic frameworks for modeling real-time applications for further schedulability and performance analyses. These two profiles do not, by and large, support all available techniques and algorithms, but rather only a representative subset of those. It is expected that more specialized profiles for schedulability and performance analyses, as well as syntheses, can later be supplied by tool vendors.

3.2 EXTENSIBLE MARKUP LANGUAGE (XML)

The Extensible Markup Language (XML) is the other technological underpinning for the proposed platform-centric SoC design method. XML provides a concrete facility for realizing the LPO concept and managing the complexity resulted from a huge database of extremely diverse LPO modules. XML also promotes information exchange which can culminate to the virtually unbounded sharing of PO member modules (POmm). Furthermore, there exists an extensive collection of public-domain and/or open-source tools for XML and other XML-related technologies that may be used to enhance its capability.

3.2.1 Introduction to Markup Languages and XML

Markup languages are all about describing the "form" of the document—that is, the way the content of the document should be interpreted [47]. A markup language is a notation for writing text with markup tags, where the tags are used to indicate the structure of the text [48]. Tags have names and attributes, and may also enclose a part of the text.

One of the most predominant markup languages these days is the Hyper-Text Markup Language (HTML). HTML is an application of the Standard Generalized Markup Language (SGML), which is a rather complicated superset of XML. HTML has a fixed set of markup tags, and is primarily used for layout on the Web. Being a hyper-text language, a HTML document can contain links to other documents, text, sound, images, and various other resources. However, it says nothing about the content of the data.

Both SGML and HTML heavily influence the development of XML [49]. XML is a semantic language that allows text to be meaningfully annotated. It is designed to separate syntax from semantics to provide a common framework for structuring information and allow tailor-made markup tags for any imaginable application domain [48]. XML also supports internationalization (Unicode) and it is platform-independent.

One of the functions of XML is the storage of data. The technologies that dominate the market for data storage are relational databases that manage traditional data types such as numbers and text [50]. XML goes beyond the boundary of traditional relational database technologies by offering the potential to process smart data, or i.e. self-describing data. XML documents are, by their nature, ideal for storing databases [51]. A standardized interface to XML data is defined through W3C's Document Object Model (DOM) [52], which provides a CORBA IDL interface [53] between applications exchanging XML data. The clearly defined format provided by XML helps make the data readily transferable to a wide range of hardware and software environments. New techniques in programming and processing data will not affect the logical structure of the document's message [51]. If more detail needs to be added to the document, the model can be updated and new markup tags added where required in the document instance. If a completely new style is required then the existing document model can be linked to the new one to provide automatic updating of document structures. For these reasons, a claim exists that XML brings about a revolution in communication and information distribution across the Web and within intranets [50].

3.2.2 Conceptual View of XML

An XML document is an ordered, labeled tree [48]. The leaf nodes of an XML tree are terminal elements that contain actual data in the form of text strings. Although a character data leaf node can be declared EMPTY, it is usually non-empty and must be non-adjacent to other leaf nodes. Other kinds of XML leaf node exist, that include:

- processing instructions—annotations for various processors,
- comments, and
- a schema declaration, i.e. Document Type Definitions (DTD) or XML-Schema.

In XML tree terminology, non-terminal elements always contain sub-elements, or child nodes, that can be grouped as sequences or choices. A sequence defines the order in which sub-elements must appear. A choice gives a list of alternatives for sub-elements. Sequences and choices can contain each other. Each non-terminal element node can be labeled with a name—often called element type—and a set of attributes, each consisting of a name and a value.

By adopting the Extended Backus-Naur Form (EBNF) for its syntax (see Section 1.4 of the official W3C XML Recommendation [55]), XML is designed to be easy to use with modern compiler tools [54]. The EBNF defines a set of rules, where every rule describes a specific fragment of syntax. A document is valid if it can be reduced to a single, specific rule, with no input left, by repeated application of the rules. XML is defined by an EBNF grammar of about 80 rules.

An XML document normally consists of three types of markup tags, the first two of which are optional:

1. An XML processing instruction identifying the version of XML being used, the way in which it is encoded, and whether it references other files or not, e,g,
   <?xml version="1.0" encoding="UTF-8" standalone="yes">

2. A schema declaration that either contains the formal Document Type Definitions (DTD) markup declaration in its internal subset (between square brackets) or references a file containing the relevant markup declaration (the external subset), e.g.:
   <!DOCTYPE poModule SYSTEM "http://www.lpo.com/dtds/poModule.dtd">

3. A fully-tagged document instance which consists of a root element, whose element type name must match that assigned as the document type name in the document type declaration (DTD), within which all other markup is nested.

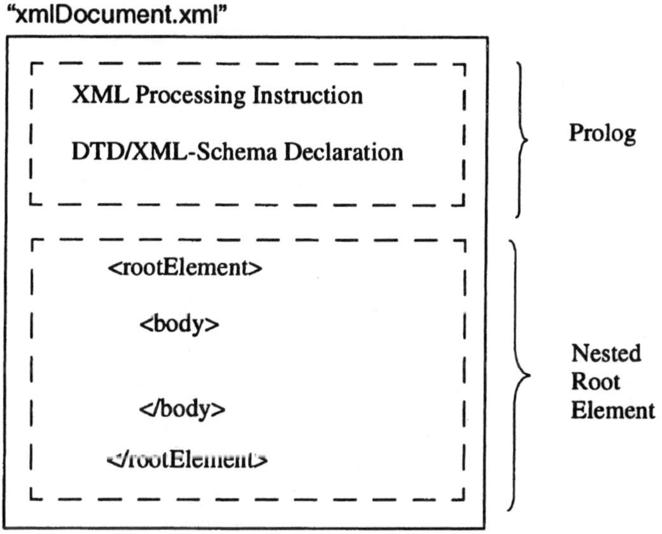

Figure 3-4. Structure of an XML document

4. An XML processing instruction identifying the version of XML being used, the way in which it is encoded, and whether it references other files or not, e,g,
   <?xml version="1.0" encoding="UTF-8" standalone="yes">

5. A schema declaration that either contains the formal Document Type Definitions (DTD) markup declaration in its internal subset (between square brackets) or references a file containing the relevant markup declaration (the external subset), e.g.:
   <!DOCTYPE poModule SYSTEM "http://www.lpo.com/dtds/poModule.dtd">

6. A fully-tagged document instance which consists of a root element, whose element type name must match that assigned as the document type name in the document type declaration (DTD), within which all other markup is nested.

The term prolog is used to describe the section where the XML processing instructions and a schema declaration may optionally appear. Figure 3.4 illustrates the XML document structure.

A well-formed XML document must contain exactly one element. This single element can be viewed as the root of the document. Elements can be nested, and attributes can be attached to them. Attribute values must be in quotes, and tags must be balanced, i.e. they must always be explicitly opened and closed. Empty element tags must either end with a /> or be explicitly closed.

The structure of an XML document is defined by a schema language. Currently the W3C XML Recommendation [55] supports two types of a schema language, namely, Document Type Definition (DTD) and XML-Schema [56]. While the XML-Schema is by far more powerful than the DTD, it is also more complicated and still is very immature. Tool support for the XML-Schema is meager. As of today, it remains to be seen whether or not the XML-Schema will be able to completely replace the DTD as stated in its objective. In implementing the LPO, only the DTD will be used for it is more stable and sufficiently powerful. The DTD to XML-Schema conversion tools are readily available. Figure 3.5 shows an example of a DTD file used to describe the structure of the PO register file.

Also used in SGML, the DTD provides a facility to specify a set of tags, the order of tags, and the attributes associated with each. A well-formed XML document that conforms to its DTD is called valid. A DTD is declared in the XML document's prolog using the !DOCTYPE tag. It is actually within the DTD that the sequence and choice grouping of non-terminal child nodes is specified. XML tags are defined using the DTD's ELEMENT tag, while the attributes associated with each XML tag are defined using ATTLIST. A DTD's terminal element can be of types parsed character data (#PCDATA), EMPTY or ANY. A terminal element declared as ANY will not be parsed; thus, it can contain sub-elements of any declared type, including character data. It is to note that an XML document is case-sensitive.

```
<!-- Define abstract type for each module -->
<!ENTITY % typeFile SYSTEM "file:///C:/Data/Thesis/XML/Test/abstractType.txt">
<!ENTITY % abstractTypeList "(%typeFile;)">

<!ELEMENT poRegfile (self, pom, blueprint, poSchema, poInfo?, polif+)>

<!-- Start with self here -->
<!ELEMENT self (id, poID, uri)>
<!ELEMENT name (#PCDATA)>
<!ELEMENT poID (#PCDATA)>
<!ELEMENT textField (#PCDATA)>
<!ELEMENT id (#PCDATA)>
<!ELEMENT uri (#PCDATA)>

<!-- the POM, PO Manager -->
<!ELEMENT pom (id, uri)>
<!-- Architecture blueprint -->
<!ELEMENT blueprint (uri)>
<!-- Location of the PO schema file -->
<!ELEMENT poSchema (uri)>
<!-- Information about the PO -->
<!ELEMENT poInfo (info)+>
<!ELEMENT info (subject, textField)>
<!ELEMENT subject (#PCDATA)>

<!-- Main Entry... consisting of toolModule and componentModule -->
<!ELEMENT polif (name, uri)>

<!-- Attribute declaration -->
<!-- This DTD uses a MIME type for the fieldType attribute -->
<!ATTLIST textField fieldType CDATA "text/plain">
<!ATTLIST textField isImported (yes | no) "no">
<!ATTLIST polif moduleKind (component | tool) #REQUIRED>
<!ATTLIST polif abstractType %abstractTypeList; #REQUIRED>
<!-- moduleID should match an id as defined in each corresponding xml file -->
<!ATTLIST polif moduleID ID #REQUIRED>
```

Figure 3-5. An example of a Document Type Definition (DTD) file. When writing the DTD as a separate file, the DTD file contains the content of the DTD declaration as nested between the square brackets within the <!DOCTYPE [DTD]> element.

3.2.3 XML Extensions and Applications

As of late 1998, the XML design effort was re-chartered under the direction of an XML Coordination Group and XML Plenary Interest Group to be carried out in five new XML working groups [57]: XML Schema Working Group, XML Fragment Working Group, XML Linking Working Group (XLink and XPointer), XML Information Set Working Group, and XML Syntax Working Group. These working groups were designed to have close liaison relationships with the W3C's Extensible Stylesheet Language (XSL) Working Group and Document Object Model (DOM) Working Group. Under such concerted collaborations, several XML-based technologies have since emerged. These technologies make XML even more powerful and attractive as a convenient tool to realize the LPO concept. Some of these XML extensions and applications that could be used to enhance the proposed platform-centric SoC design method include:

- Namespaces: The use of namespace helps avoid name clashes when the same tag name is used in different contexts. Namespaces can be defined in any element, and appear as a prefix before an element or an attribute name, separated by a colon. Their scope is the element in which they are defined. Unfortunately, namespaces and DTDs do not work well together [49]; they bear no special meaning for the DTD. The DTD views a namespace as an integral part of a tag name without differentiating it from the local name. To better harness the power of namespaces, XML-Schema should be used.

- Addressing and linking: XML extends HTML's linking capabilities with three supporting languages: Xlink [58] which describes how two documents can be linked; XPointer [59] which enables addressing individual parts of an XML document; and, XPath [60] which is an underlying technology for XPointer for specifying location paths. These technologies, together with the namespace technology, enable XML documents, e.g. the POmm, to integrate very well into the World Wide Web. As a result, data exchanges are promoted in such a way that can ultimately lead to one big database that spans the entire Internet and intranet spaces.

- WWW Resources: The Resource Description Framework (RDF) [61] is an XML application that works with metadata, i.e. data that describes data; it expresses XML data in the canonical XML format for ease of identification. RDF is a language that describes WWW resources, such as title, author and modification date of a Web page, as well as anything that can be identified on the Web [62]. It provides a common framework for expressing WWW information in such a way that it can be exchanged between applications without loss of meaning. One widely used RDF today is the Dublin Core, which is seen by many as a way of standardizing RDF for Web resources [47]. As the LPO can also span the WWW domain, the Dublin Core may be used to ease the POmm identification process. More information about the Dublin Core can be found at http://www.purl.org/dc. Caveat: Many of these web-based URLs are likely to be updated frequently.

Of these XML extensions and applications, XPath plays a major role in an efficient implementation of the LPO. Because the LPO is composed mainly of XML data, XPath becomes an obvious choice for data querying and retrieving operations. In addition, other XML-based technologies such as XPointer, XQuery [72] and XSL/XSLT [71], that can potentially contribute to the success of the LPO, also base their implementations on or around XPath. The applicable usages of these technologies may involve using (1) XPointer to access remote XML content, (2) XQuery to process XML data query, or (3) XSL/XSLT to automatically render XML data into another format more suitable for human understanding.

3.2.3.1 XPath

XPath is a syntax for defining parts of an XML document [73]. The standard also defines a library of functions to handle strings, numbers and Boolean expressions. XPath perceives an XML document as a tree view of nodes. It uses location path expressions similar to the traditional file path to identify nodes in an XML document. A location path expression results in a node-set that matches the path.

A location path consists of one or more location steps, separated by a slash (/), and it can be absolute or relative. An absolute location path starts with a slash and a relative location path does not. The location steps are evaluated in order, one at a time, from left to right. Each step is evaluated against the nodes in the current node-set. If the location path is absolute, the current node-set consists of the root node. If the location path is relative, the current node-set consists of the node where the expression is being used. Location steps consist of an axis, a node test and zero or more predicates. The syntax for a location step takes the form: axisname::nodetest[predicate]. For example, in the expression child::po[last()], child is the name of the axis, po is the node test and [last()] is a predicate. XPath 1.0 supports only four expression types and seven node types. The expression types consist of node-set, string, number, and boolean; whereas, the node types encompass document, element, attribute, comment, text, namespace and processing instructions.

Table 3-3. XPath syntax abbreviation

Expressions	Abbreviation
self::node()	.
parent::node()	..
child::childname	childname
attribute::childname	@childname
/descendant-or-self::node()/	//

[position()=3]	[3]
[position()=last()]	[last()]

Table 3-4. Examples of the XPath expressions

Path	Description
/lpoRegfile/self/*	returns all children of the *self* node, i.e. *name, id, uri, textField*
@*	returns all attributes of the context node
string(//id/.)	returns string value of the *id* node, i.e. *lrf001*
//attribute::isImported	returns the *isImported* attribute in the document, i.e. *isImported = "no"*
/descendant::po[1]	returns the first *po* node in the document
string(//@fieldType)	returns the string value of the *fieldType* attribute in the document, i.e. *text/plain*
po[name and uri]	returns all the *po* nodes within the context that contain both *name* and *uri* elements
//self/textField[@isImported = "no"]	returns all *textField* nodes that have the *isImported* attribute values set to "no"
count(//uri)	returns number of the *uri* node appearances, i.e. 3

To make XPath expressions easier to use, abbreviations are defined for frequently used XPath syntax. Table 3.3 lists these abbreviations. Table 3.4 demonstrates the use of some XPath expressions on the LPO register file below.

```
<?xml version="1.0"?>
<!DOCTYPE lpoRegfile SYSTEM
            "file:///C:/Data/Book/XML/Test/lpoRegfile.dtd">
<lpoRegfile>
  <self>
    <name>Example LPO</name>
    <id>lrf001</id>
    <uri>file:///C:/Data/Book/XML/Test/lpoRegfile.xml</uri>
    <textField>This is a description of this lpoRegfile</textField>
  </self>
  <po>
    <name>po-NiOS</name>
    <uri>file:///C:/Data/Book/XML/Test/poRegfile_nios.xml</uri>
  </po>
  <po>
    <name>po-OMAP</name>
    <uri>file:///C:/Data/Book/XML/Test/poRegfile_omap.xml</uri>
  </po>
</lpoRegfile>
```

A number of public-domain tools exist for XPath. For example, the Apache's Xalan [74, 75] contains an XPath engine that can be used with

C++ or Java. The Java-based Jaxen [76] and Microsoft's MSXML [77] provide some alternatives. The examples presented in Table 3.4 are evaluated using MSXML 3.0. Complete reference to XPath can be found at the W3C website [60]. ZVON [78] and W3Schools [73] also contain excellent resources on XPath and other XML-related topics. As it is beyond the scope of this book to cover the whole spectrum of XML technologies in detail, interested readers should consult appropriate references for further information. Chapter 4, in addition, will discuss how each of the XML technologies presented above fit into the LPO implementation scheme.

Chapter 4

LIBRARY OF PLATFORM OBJECTS

4. INTRODUCTION

This chapter is dedicated to a detailed discussion of the XML specification for realizing a library of platform objects (LPO) for the proposed platform-centric CMF methodology. It starts by describing desirable LPO characteristics from the conceptual domain viewpoint. From within this domain, requirements for the LPO are identified. Thereafter, the chapter maps these requirements onto the XML domain, and defines the XML specification, as well as identifies the roles that are expected from the system developer, the platform object member modules (POmm) developer, and the POmm themselves. It provides a general but precise guideline on how an LPO should be implemented. Requirement levels are indicated by keywords according to the guideline furnished by the Internet Engineering Task Force (IETF)'s RTF2119 [63], which is also summarized in Table 4.1.

4.1 CONCEPTUAL VIEWPOINT

This section looks at an LPO analytically, and identifies characteristics that are desirable for the proposed platform-centric SoC design method. Once identified, these characteristics serve as the basis for deriving the XML specification. In general, an LPO behaves like a data warehouse for the developer—it permits data to be stored, inventoried, searched, and retrieved. The specification addresses such requirements, as well as ensures compatibility among data from different providers, and among every individual LPO and the proposed approach.

Table 4-1. Summary of the requirement levels as specified in IETF's RFC2119

Keywords	Meanings
must, required, shall	an absolute requirement
must not, shall not	an absolute prohibition
should, recommended	there may exist valid reasons in particular circumstances to ignore a particular item
should not, not recommended	there may exist valid reasons in particular circumstances when the particular behavior is acceptable or even useful
may, optional	truly optional

Definition 4.1. Library of Platform Objects (LPO)
A collection of pre-designed, pre-characterized system platforms that further meet a governing set of rules and requirements specific to the proposed platform-centric SoC design method.

Definition 4.2. Platform Object (PO)
A configurable/reconfigurable, pre-designed, pre-characterized system platform that (1) supports *fast* and *correct* system construction via a set of well-defined communication infrastructure, and (2) carries with it a set of compatible *platform object member modules (POmm).*

4.1.1 LPO in Principle

AN LPO is a distributed and scalable database. As such, it is only appropriate to, first and foremost, identify with these principal LPO characteristics that make the concept novel and utility for the proposed platform-centric SoC design method.

Database-wise, from the platform-centric approach's point of view, an LPO involves three main categories of data:

- *Component information.* Data in this category conveys various information about component characteristics, as well as component-specific knowledge for the very purpose of assisting the developer in efficiently deploying its self. The data shall include, at the minimum, the component identification and its whereabouts, the corresponding UML model, the characteristics similar to those appeared in a databook, and potentially the availability of component-specific software routines, as well as any possible tool(s) association. Knowledge-based information, such as design guideline, user manual, official specification, etc., is also expected.

- *Tool information.* Because platforms are often designed with certain application domains in mind, full potential of a platform-based design such as the proposed approach can usually be benefited when a platform is accompanied by a specially-tailored tool suite. For this reason, tools, and not just components, are integral to the robustness of the platform-centric approach. To treat tools and components equally, the method views tools also as LPO modules that can convey information, just like their component counterparts. Information held by tool modules is not as diverse as that of component modules, and shall include information about their availability and whereabouts, as well as knowledge-based information such as user guide and manuals.

- *Platform information.* A platform often is configurable using both hardware and software components. In addition, because of its specialized nature, it is suitable for a platform provider to also supply a template that can serve as a guideline for the developer when trying to construct a desirable target architecture from the platform. As such, various characteristics related to this template can be collected and grouped, and the template placed in an LPO.

An LPO is scalable. Once constructed, it is open to an addition of new modules and/or a removal of current ones—given that such an addition or removal does not conflict with the requirements for its existence. It is perceived that an LPO could behave like a distributed database, with platforms (PO) and platform components (POmm) physically residing across the Internet and the developer only locally maintains the logical representations of these entities. Owing largely to the Internet technologies, such "distributedness" and scalability potentially renders better manageablility, maintainability and upgradability for an LPO—a feature that could very well be attractive for both the system developer and the platform and component providers. Under such a scenario, the system developer could be completely up-to-date with currently available platform technologies and components; whereas, the platform and component providers could be in total control of the services such as component usage, upgrade, and maintenance.

4.1.2 Identity

One of the questions with an LPO behaving like a distributed database is, "whose responsibility is it to first initiate the existence of an LPO?" Because an LPO is primarily a database of platforms whose existences further induce the existence of platform components, it should be the platform providers that are responsible for the initiation. When a platform is installed, it shall check to see if an LPO exists that can be identified. If not, it creates one and makes it known publicly such that subsequent insertion of platforms will not have to perform the same task redundantly.

The ability to check an existence of an LPO at the time of platform installation implies that there must locally exist a mechanism that keeps track of platform availability as well as any relevant information necessary for identifying and locating platforms. This mechanism represents a logical LPO space that provides the mapping to their physical counterparts during the development process.

Definition 4.3. Logical LPO space
A logical representation of an LPO that resides locally, where the developer has complete access to, and providing links to its physical equivalents

Definitiion 4.4. Physical LPO space
A physical LPO space defines a "locationally" unrestricted domain where an LPO and its data may reside. This spatial domain is confined by the Internet boundary, and includes any local space as well. As such, it is a superset of a logical LPO space.

4.1.3 Scalability

AN LPO grows or shrinks dependently on an attachment or removal of LPO modules, i.e. platforms and/or platform components. An attachment of an LPO module onto an LPO requires, as a prerequisite, that the attaching module physically exists. Its whereabouts are "locationally" unrestricted—it may be locally present, or remotely accessible via the Internet; it attaches itself onto an LPO by creating a logical instance of itself and imprinting that instance onto the logical LPO space such that a trace to its physical counterpart is possible and complete, i.e. trace information from the logical LPO space can bring about complete access to information stored remotely at the physical LPO space. An act of removal is essentially the reverse concept of an attachment. However, it is only necessary to remove a logical module from the locally located logical LPO space—effectively detaching a link between the physical counterpart and the logical LPO. To promote scalability, physical LPO modules can also make their presences known and identifiable to Internet search engines.

An installer or an uninstaller can be useful for attaching and removing logical LPO modules, be they platforms or components. The use of an installer/uninstaller does also offer another benefit; it makes an LPO more user-friendly and attractive, resulting in a higher frequency of LPO utilization. In "The Selfish Class" [69], the authors suggest through the "Work Out Of The Box" pattern that a higher rate of software object utilization usually results in a better rate of survival. This principle applies to the use of default arguments as well.

4.1.4 Operations

LPO-related operations, chiefly *read data* and *process data*, are performed by LPO tools. Each LPO tool must be able to read needed data from the XMI inputs and/or the platform components, and then process them according to its intended task.

AN LPO may also furnish the developer with a user-interface tool that makes interactions between the developer and the LPO efficient. This tool shall support the Create/Read/Update/Delete (CRUD) operations, similar to the basic operations found in most Object Oriented and Query languages, to help manage local modules in the logical LPO space. Moreover, because of the information-rich nature of an LPO, the user-interface tool should also be able to process data contents and present them in a human-readable format, in much the same way as web browsers do.

By observing how an LPO comes into existence, it is obvious that each platform must also carry with it an user-interface tool. Since it is possible that an LPO can host more than one platform, there can very well be more than one user-interface tool within an LPO—each one comes with and belongs to each platform, and can be tailored specifically to suit the platform's characteristics. The platform that first initiates the LPO creation shall lend the service of its user-interface tool to help manage modules in the locally-located logical LPO space.

4.1.5 Interactions

The platform-centric SoC design method relies heavily on two mechanisms, namely, the UML and an LPO. These mechanisms belong to two different application domains; UML applications are allowed to request cross-domain services from an LPO. Because of its nativeness to UML and its being based on XML just like the LPO, it is envisaged that the XML Metadata Interchange format (XMI) could be a natural choice that provides a standard means for representing inter-domain communications. Operations and components provided by an LPO constitute the processing power and development resources required to drive the design flow for applications modeled in the UML domain.

Utilizations of LPO components during the development process, either in the context of target architecture construction or application design, are achieved through the use of the UML's Package which permits these components to be imported into the UML application domain for further reuse. As such, it is imperative that each LPO component provide a UML package whose contents contain links to itself in an LPO, and may describe the behaviors and characteristics that will promote the reuse of its own self in the UML domain.

4.2 XML VIEWPOINT

This section describes an LPO from an implementation point of view, where XML promises to be a convenient, yet effective, tool for realizing the library. To begin, the mapping from the conceptual viewpoint to the XML equivalents are presented. The actual implementation of an LPO then ensues that lays out the general structure, as well as the structures of the schema documents, including relevant tag definitions, that constitute the physical artifacts of the LPO.

4.2.1 Mapping of Conceptual LPO to XML Equivalents

Not all the LPO characteristics from the conceptual domain can be explicitly mapped one-to-one into the XML domain. However, by imposing certain rules to those with no direct mapping, XML could suffice to completely realize all the characteristics of the LPO.

The logical LPO space as defined in Definition 4.3 can be regarded as a locally-located database that logically represents a larger set of real data whose locations could be anywhere in the physical LPO domain. When physical LPO modules are attached to or removed from an LPO, this local database is modified to reflect such changes. In the XML domain, such a local database manifests itself as a register file.

Definition 4.5. Register File
An editable document that holds information about the identity and whereabouts of either the platform objects (PO) or PO member modules (POmm), as defined by the lpoRegfile and poRegfile schemas, respectively.

4.2.1.1 LPO Register File

The existence of the LPO register file signifies the existence of an LPO. It is created when the first platform object is installed. Each subsequent PO installation updates the LPO register file in order to declare its presence in the LPO (via the logical LPO space).

Axiom 4.1. For the Library of Platform Objects L, L exists if and only if the corresponding LPO register file R_L exists.

Axiom 4.2. Let S_{RP} be a set of PO register files in L. R_L exists if and only if S_{RP} is a non-empty set.

Axiom 4.3. From Axioms 4.1 and 4.2, it follows that L exists if and only if S_{RP} is a non-empty set.

4.2.1.2 PO Register File

The existence of the PO register file signifies the existence of a platform object (PO). It is created when the platform object (PO) is installed. Each subsequent POmm installation updates the PO register file in order to declare its presence in the PO.

Axiom 4.4. For the platform object P, P exists if and only if the corresponding PO register file RP exists

4.2.1.3 Auxiliary Information

Because XML that implements an LPO depends on schema documents to define tag syntax and semantics, present also as part of the library is the fourth data category—the auxiliary information.

In addition to schema documents, the category actually encompasses anything at all that cannot be classified into components, tools or platform information, but are present either as requirements for proper functioning of an LPO or as auxiliary entities for efficiency gains. Other auxiliary information may include user-defined definitions to be used by the schema documents, or run-time configuration files used by the user-interface software, as well as the register files.

4.2.1.4 Structure of LPO

Figure 4.1 illustrates the structural organization of an LPO. UML's Aggregation notations depicted as straight lines with hollow diamonds, are used to represent hierarchical containment relationships, where a module on the hollow diamond end contains the other module. Straight lines connecting modules simply show Association relationships among them.

Definition 4.6. Platform Object Member Module (POmm)
A member of a particular platform object that is used to design and construct a system. Four kinds of module exist in a platform object, corresponding to four LPO data categories: component, tool, architecture blueprint (or platform information), and auxiliary.

Definition 4.7. Architecture Blueprint (AB)
Sometimes referred to only as blueprint, it is an abstract, logical view of the PO architecture(s). It corresponds to the template notion introduced in Section 5.1.1.

Definition 4.8. Platform Object Logical Interface (POlif)
A collective term used to describe the POmm modules, i.e. POmm/tools and POmm/components, that provide a logical gateway to the corresponding physical entities existing elsewhere in the physical LPO space.

Definition 4.9. Platform Object Manager (POM)
A software managing tool whose main task is to provide for the system developer a user interface to the member modules of the same PO it belongs to. It corresponds to the concept of user-interface tool described in Section 5.1.4.

4.2.1.5 Sole Point of Access

As shown in Figure 4.1, a POM should be capable of providing the sole point of access to the PO domain it belongs to. By further elaborating on POM functionalities, a seamlessly unified tool environment for the proposed approach could possibly be attained. Figure 4.1 also roughly shows the interactions between the LPO and its environment, i.e. system developers and POmm providers.

4.2.1.6 Tag Syntax and Semantics

An LPO can become more efficient if its XML tag syntax and semantics possess a certain degree of flexibility that enable them to be tailored specifically for each platform. Nonetheless, for an LPO to function for the proposed platform-centric SoC design method, a common set of LPO tag syntax and semantics must be agreed upon.

To promote (1) a possible sharing of multi-PO POlifs, and (2) the use of a common POM to perform basic services for all the POs, all register files in the LPO shall resort to one common set of XML tag syntax and semantics.

Also, all XML documents within each PO domain—specifically, the POlif domain—must utilize the same syntax and semantics. In addition, all XML and XMI documents must be valid when checked against appropriate schema files.

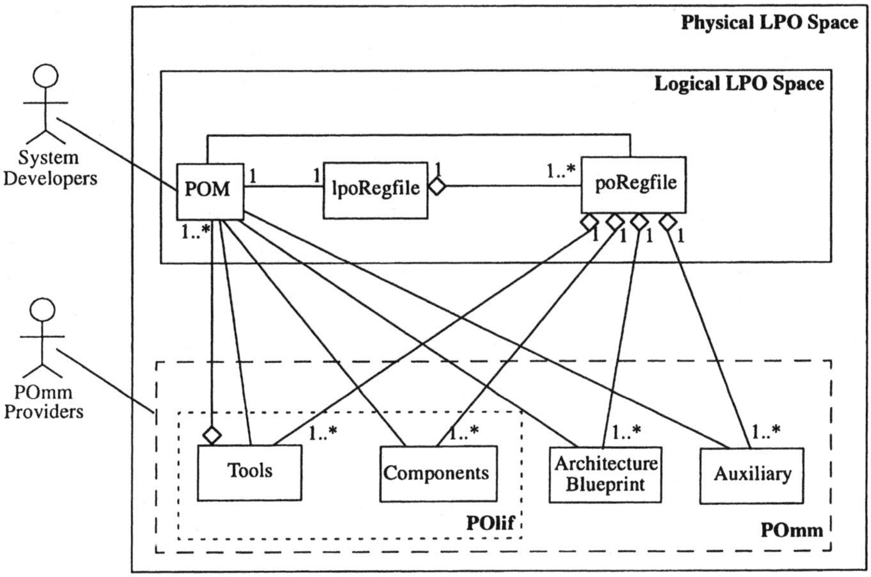

Figure 4-1. Structural organization of the LPO

Let P be a platform object in an LPO L. Let $D_u \in P$ be a set of XMI documents, $D_p \in P$ be a set of XML documents in the POlif, $S \in P$ be a set of the schema documents associated with each elements in D_p, and μ be an official OMG's XMI schema. Also, let *isValid(x, y)* be a function that returns TRUE if an XML document *x* is *valid* when checked against a schema *y*, and returns FALSE otherwise.

Axiom 4.5. A platform object P can only effectively contribute to the proposed platform-centric SoC design method if, (1) $\forall d_u \in D_u$, *isValid*(d_u, μ) *is TRUE, and (2)* $\forall s \in S$ *and* $\forall d_p \in D_p$, *isValid*(d_p, s) *is TRUE.*

Furthermore, for any active PO, there must exist at least three schema documents at all times: the polif and poRegfile schemas, and the XMI schema as defined in the UML [24].

4.2.1.7 Platform Objects

Platform objects are designed off-cycle. By so doing, the enhanced flexibility offered by the proposed approach, or any platform-based design, is somewhat compromised. In return, however, the platform-centric approach achieves greater potential to attain the feasible design correctly and more quickly by pre-designing, pre-characterizing certain aspects of the system, as well as providing guidance both in the forms of constraints and related information so that the system developer can make better design decisions at all stages of the design.

A PO enters the LPO by updating the LPO register file to record its presence. Upon entry, various tasks may also need to be carried out and proper values configured. These configuration tasks may involve:

- determining and selecting existing design tools, and their execution paths,
- determining design tools that must be installed,
- determining the type of a computing platform the POM will be deployed,
- determining if an LPO already exists.

An installer and/or a clear installation instruction should always be provided to ease the installation process.

> *Axiom 4.6. Let L be an LPO, and n be the number of POs in L. Then if there exists L, then L must contain at least one PO as its library member, i.e. exist(L)* $\rightarrow$ *(*$n \geq 1$*).*

At installation time, the PO checks for the existence of the LPO register file to determine whether or not an LPO exists. If it finds no LPO register file, it creates one and updates the file to declare its presence. Then it creates its own PO register file and performs necessary updates to reflect the existence of the POmm modules that are vital to its existence. These POmm modules include:

- The architecture blueprint (AB), and POmm/components that build it,
- Associated POmm/tools, including the PO managing software (POM),
- Schema documents

- Knowledge-based information

It is expected that most of the PO for the proposed approach will come as a result of modifying existing system platforms. Two broad classes of system platforms are likely to become common in the LPO:

Full-application platform, which allows full applications to be built on top of hardware and software architectures. In general, the blueprint will consist of a processor and a communication infrastructure. The POmm/components are composed mainly of application-specific blocks that will probably share POmm/tools. Examples include Philips' Nexperia [79] and TI's OMAP multimedia platform [80].

Fully programmable platform, which typically consists of a FPGA and a processor core. Communication infrastructure is often synthesized along with the core, on an as-needed basis, during configuration. Examples include Altera's NiOS [81], Quicklogic's QuickMIPS [82], and Xilinx's Virtex-II Pro [83].

A choice of PO communication infrastructure often plays a vital role in achieving high performance and great flexibility. Besides easy integration, desirable I/O subsystems should also take into account scalability and parametrizability. CAN [85], FlexRay [86] and I2C [87] represent a subset of current cutting edge embedded system I/O technologies that can potentially be used to build a platform. An I/O subsystem that allows its power to be configured [88] is also an attractive choice. In any case, proper documentation should always be exercised.

4.2.1.8 Architecture Blueprint (AB)

Because a PO either comprises of a family of processors, or is fully programmable, it can be affiliated with more than one target architecture. A blueprint is an abstract and logical view of these architectures.

Working with a blueprint instead of the physical model consisting of UML nodes and components may be more appealing to the system developer. It is a convenient means to represent a set of all possible target architectures attainable per platform. Moreover, it provides a simple yet powerful mechanism for dealing with the issue of hardware implementation that, otherwise, would not be satisfactorily addressed were the developer to adhere to the UML physical model. As will be shown later in Section 5.6, homomorphic mapping between the logical model of the chosen target architecture derived from an architecture blueprint and the physical model, if it were to be used, is possible, and thus, furnishing a proof that the two views are, in fact, equivalent and interchangeable.

At the very minimum, an architecture blueprint must consist of an abstract representation of a processor and a communication infrastructure, the latter of which may further consist of one or more I/O subsystem abstracts. Given the current trend in platform technologies, an architecture blueprint that represents a family of processors or multi-processors, rather than a traditional uniprocessor, will not be uncommon. A processor may contain one or more internal storage elements and/or have external storage elements as an additional AB requirement.

An architecture blueprint is an integral part of the proposed approach; it exists necessarily for every PO in the library. Because there always exists at least one processing unit and one I/O component for every PO, it could be further deduced that if a blueprint exists, the POmm/components that are used to construct the concrete platform architecture must also exist.

Axiom 4.7. Let P be a platform object and $S_{AB} \in P$ be a set of the POmm/components that can be instantiated into the blueprint to implement the concrete platform architecture. Then exist(P) $\rightarrow$ ($S_{AB} \neq \varnothing$).

4.2.1.9 Platform Object Logical Interface (POlif)

By definition (Definition 4.8), POlif is a collective term used to describe POmm/tools and POmm/components. POlif modules are the core database of the LPO. Like others, POlif modules are self-descriptive. When implemented using XML, the POlif shall carry its own schema document to differentiate itself from the auxiliary domain of the register files. They shall also permit keyword description of themselves to aid search engines.

A POmm/tool is a logical interface module to the corresponding tool that may physically exist anywhere in the physical LPO space. It contains essential information about itself, especially its identity, and resource locations, that could be configured during the installation process.

A POmm/component, on the other hand, carries a much heavier load of data than a POmm/tool does. In addition to the information about its identity and resource locations, it contains information regarding its characteristics, UML representation, and possibly HW-dependent software routines, as well as tool associations. For each POmm/component characteristic, information about name, type, value, and unit shall always be recorded.

4.2.1.10 Resource Locator

All resources in the LPO should be specified using one common format. It is recommended that the LPO follow the Universal Resource Identifier (URI) format [89] when specifying resource locations in XML documents—the obvious reason for it being the compatibility with the Internet standard. The most common form of the URI is the Universal Resource Locator (URL).

4.2.1.11 Platform Object Manager (POM)

A POM is a software managing tool whose main task is to provide for the developer an interface to the member modules of the same PO it belongs to. A POM supports functionalities similar to those of the Facade object [68] which provides a unified interface to a set of interfaces in a subsystem. Like the Facade object, a POM should be able to delegate the design, either parts or whole, to appropriate POmm modules. If all members of the LPO adhere to the same tag semantics, a POM can extend its services to encompass all LPO members, including the LPO register file, and all PO register files. Let P be a platform object that contains a set of POmm/tools, P_T. And let $T_{POM} \in P$ be a POM.

Axiom 4.8. $exist(P) \rightarrow (exist(T_{POM}) \wedge (T_{POM} \in P_T))$.

POM's basic operations involve extracting and processing XML-based data. To extract data, a POM may simply make use of existing XPath engines as convenient tools. These tools typically take an XPath expression as the argument, and often provide the programming interfaces to the popular programming languages like C++ and Java [74, 75, 76]. Because of the facility and functionalities offered by XML and XPath, respectively, data embedded in a POmm module are readily accessible so far as their semantics are clearly understood. A POM can construct, as well as reconstruct, any XML tree from the associative schema document. A path expression for each element and attribute can then be acquired simply by traversing the XML tree.

The other primary POM operation is processing the extracted data. In order for the system developer to make use of the available data to the fullest, a POM should also behave as a user interface that can perform such tasks as displaying data, and relevant information in a user-friendly format, gathering input information for POmm/tools, searching and locating POmm modules, etc. At the minimum, a POM should support the following operations:

Searching and fetching LPO modules, e.g. POlif, the architecture blueprint

Displaying and formatting data, e.g. displaying Help pages, listing available POmm/components. XSL/XSLT [71] may be quite useful for this purpose.

Providing links, and allowing user selections so that such tasks as accessing remote content, activating POmm/tools, etc., may be implemented. XML applications such as XLink [58] and XPointer [59] could be useful.

Easing POmm/tool usage. A POM may (1) prepare a batch file, (2) provide step-by-step instructions on how to run a POmm/tool on a particular set of inputs, and/or (3) create tool menus that link tool and inputs together, and that permit tool activation. A POM may also define a unified environment that allows POmm/tools to interact, e.g. the OMG's CORBA IDL [53] that permits the compliant objects to communicate through an object broker.

4.2.2 Implementation

As the *tree* view structure of XML documents is easier to understand, it may be adopted as a means to describe the implementation of the LPO. Consequently, the mapping of a tree structure to an equivalent schema document is quite simple and intuitive.

The UML's Class diagrams can be used to model the tree structure. The root element of an XML tree resides solely at the topmost hierarchical level. Child elements that branch out of their parent are connected to the parent through the UML aggregation, with the hollow diamond attached to the parent. An order of child elements is significant and is mapped from left to right onto the schema document. Leaf nodes in the tree structure represent the XML terminal nodes that contain strings of type #PCDATA. UML attributes map into required XML attributes; whereas, UML multiplicity becomes the equivalents in XML.

4.2.2.1 LPO Register File

Figure 4.2 depicts the hierarchical structure of the LPO register file (lpoRegfile.dtd). Due to dynamic nature of resources, a fail-safe principle of redundancy is exercised to ensure consistency and reliability. Consequently, multiple elements may exist solely to identify a single resource. A description of each element is listed below:

Name	Type	Multiplicity	Description
lpoRegfile	Root	1	Signifies existence of LPO
self	Element	1	Self identification
po	Element	1..*	Link to platform objects
searchkey	Element	0..1	Relevant keywords that can identify self
name	Leaf	1	Self ID by name
id	Leaf	1	Self ID by special identification
uri	Leaf	1	Self ID by location
textField	Leaf	0..1	Knowledge-based information
key	Leaf	1..*	Keyword string

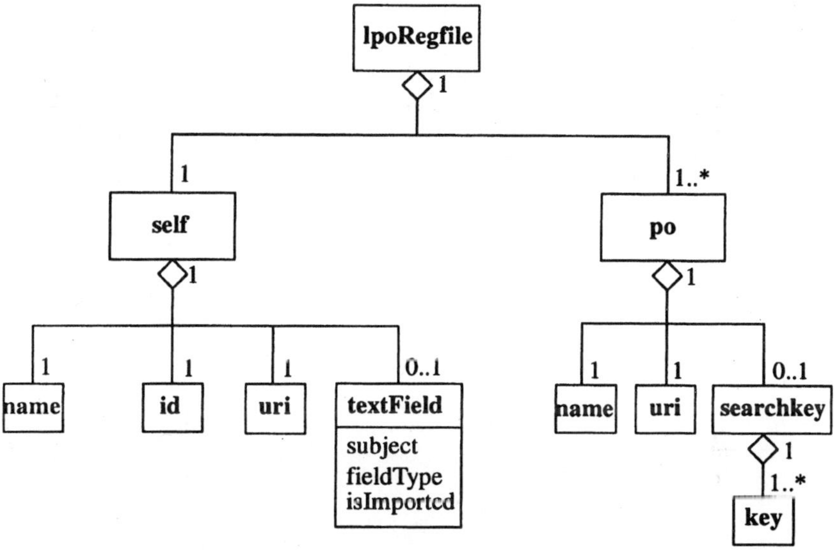

Figure 4-2. Hierarchical structure of the **lpoRegfile.dtd**

Attributes are listed as follows:

Name	Type	Base	Description
subject	CDATA	textField	Subject of information
fieldType	CDATA	textField	Expected data format
isImported	Enumeration: "yes" or "no"	textField	Specifies if the content contains link to an imported document

4.2.2.2 PO Register File

Figure 4.3 depicts the hierarchical structure of the PO register file (poRegfile.dtd). Same XML element names share syntax and semantics; thus, only descriptions of new elements are presented.

Name	Type	Multiplicity	Description
poRegfile	Root	1	Signifies existence of PO
pom	Element	1	Link to POM
blueprint	Element	1..*	Link to architecture blueprint(s)
poSchema	Element	1	Link to PO schema document
polif	Element	1..*	Link to POmm
poID	Leaf	1	Reference to the PO to which the register file belongs

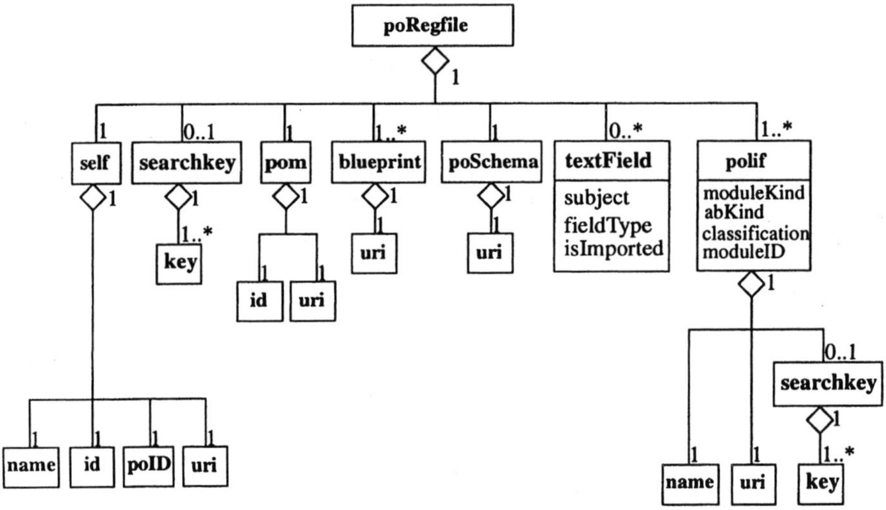

Figure 4-3. Hierarchical structure of the **poRegfile.dtd**

Attributes are listed as follows:

Name	Type	Base	Description
moduleKind	Enumeration: {"component", "tool"}	polif	Classifies itself to be either *component* or *tool*
abKind	Enumeration: {"pru", "iu", "diu", "ifu", "mu", "clock", "timer"}	polif	Classifies itself to be one of the blueprint types (see Section 5.6.1 for detail).
classification	Enumeration: *PO-dependent*	polif	User-defined category of the module

moduleID	CDATA	polif	Reference to a POmm via ID. Provides a security measure to name and uri references.

4.2.2.3 POlif

As evident by prior discussions, the POlif constitutes the core of an LPO database. POlif modules, POmm/tools and POmm/components, contain information necessary for characterizing themselves to be used with the platform-centric SoC design method. To promote scalability, each POlif module is associated with a unique XML document, which is defined by a schema document illustrated as the tree structure in Figures 4.4 - 4.8. In Figure 4.4, most elements are re-used. Those that need be defined are:

Name	Type	Multiplicity	Description
polif	Root	1	Signifies existence of a POmm
selfURI	Element	1	Possible location that it may reside
physicalURI	Element	1	Possible locations the corresponding physical module may reside
installerURI	Element	0..1	Link to an installer
uninstallerURI	Element	0..1	Link to an uninstaller
componentDomain	Element	0..1	Compartment for information about POmm/component. It is not used if a POlif is of type *tool.*
associatedTools	Element	0..1	Specifies possible association between a POmm/component and a POmm/tool(s)
uml	Element	1	UML representation of the module
functions	Element	0..*	Supplies information, if any, about hardware-dependent software routines
characteristics	Element	0..1	Contains databook information of the module

Figure 4.5 depicts a detailed structure of the associatedTools element. This tag element only supports a simple association between a POmm/component and one or more POmm/tools. It is expected that, when given the POmm/component identity in the LPO, the associated POmm/tool possesses the knowledge on how to process it. Subelements that need to be defined for the associatedTools elements are:

Name	Type	Multiplicity	Description
defaultToolURI	Leaf	1	Default tool by URI. Cannot co-exist with defaultToolID.
defaultToolID	Leaf	1	Default tool by ID. Cannot co-exist with defaultToolURI.
aTool	Element	1..*	Identity of each associated tool

Figure 4.6 shows the detailed structure of the uml element. Sub-elements that have not yet been defined are described below.

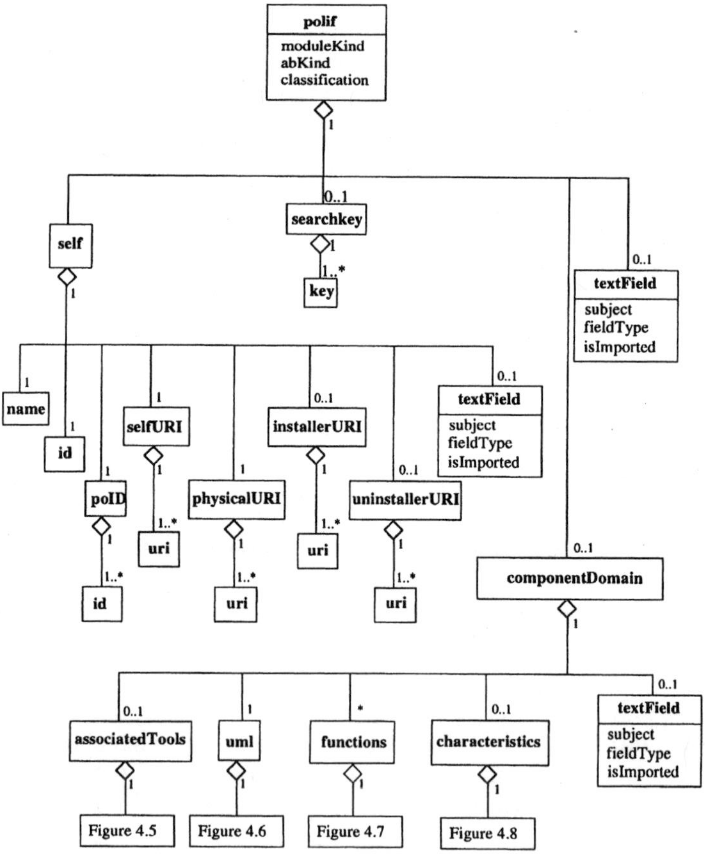

Figure 4-4. Hierarchical structure of the **polif.dtd**

Name	Type	Multiplicity	Description
defaultToolURI	Leaf	1	Default tool by URI. Cannot co-exist with defaultToolID.
defaultToolID	Leaf	1	Default tool by ID. Cannot co-exist with defaultToolURI.
aTool	Element	1..*	Identity of each associated tool

Figure 4.6 shows the detailed structure of the uml element. Sub-elements that have not yet been defined are described below.

Name	Type	Multiplicity	Description
import	Element	0..1	Reuse mechanism that allows UML packages to be imported
autoConfig	Element	0..*	Compartment that holds pre-configured values for UML parameters
config	Element	1..*	Pre-configured UML parameter values
forPOID	Leaf	0..1	ID of platform object that these pre-configured values are applicable for

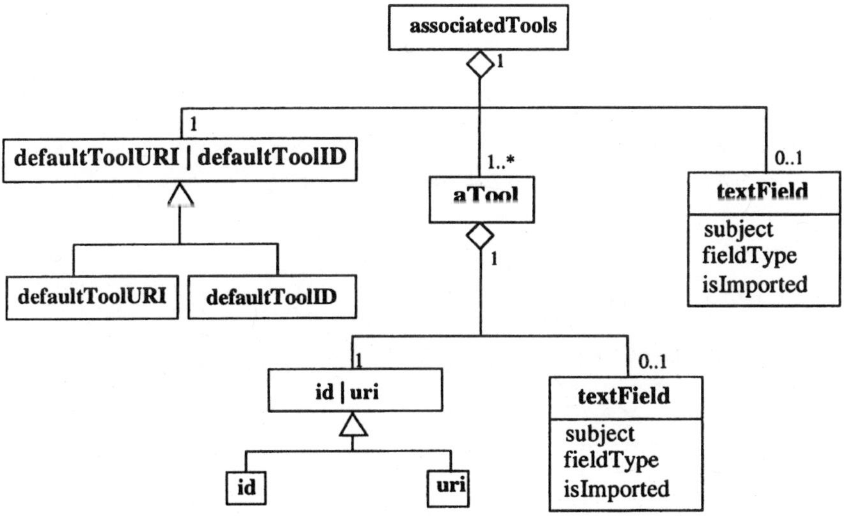

Figure 4-5. Detailed hierarchical structure of the **associatedTools** element

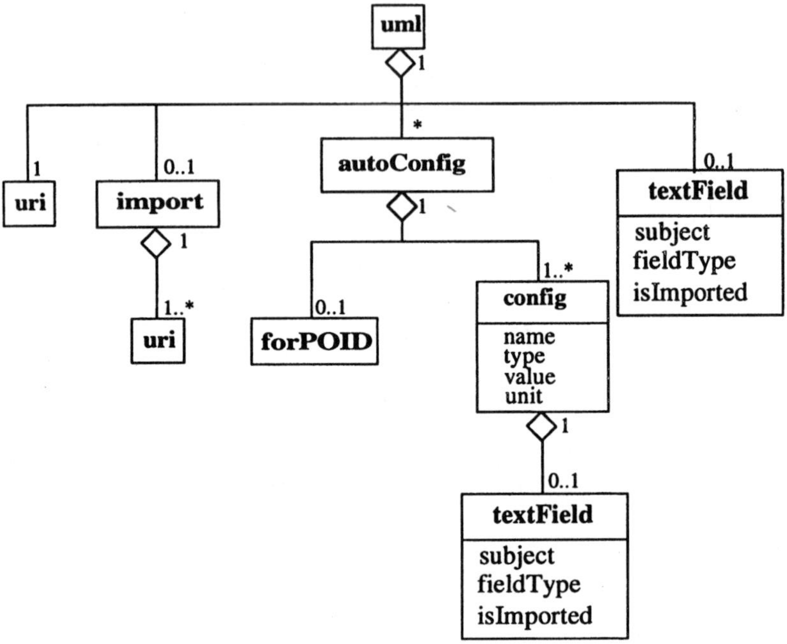

Figure 4-6. Detailed hierarchical structure of the **uml** element

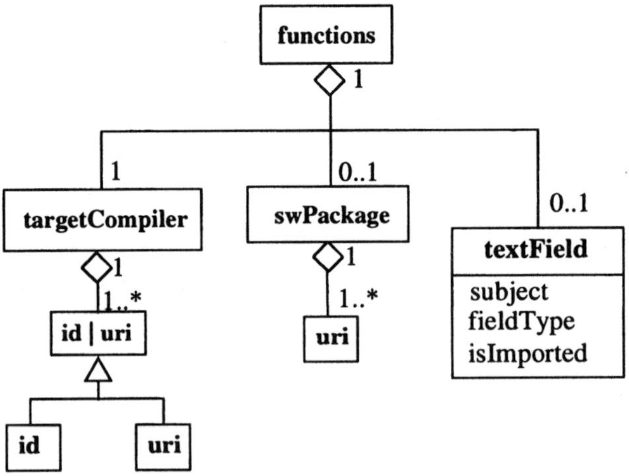

Figure 4-7. Detailed hierarchical structure of the functions element

The config subelement also have attributes associated with it. They are:

Name	Type	Base	Description
name	Enumeration: *PO-dependent*	config	Name of config data
type	Enumeration: *PO-dependent*	config	Pre-defined type of config data
value	CDATA	config	Value of config data
unit	CDATA	config	Unit of config data

The construct of the config element, as well as the preDefined, and userDefined characteristics to be discussed later, adapts the PROPERTY pattern [90] in such a way that a number of such tags that can be cataloged in an XML document can vary without any changes to the schema document. This book expects that POlif providers would eventually agree upon a comprehensive set of predefined types and units, as well as UML parameters and component characteristics. Then fairly standard *enumeration* types can be defined in the schema document.

Figures 4.7 and 4.8 depict the tree structures of the functions and characteristics elements, respectively. The functions element handles information pertinent to hardware-dependent software routines. Within it, the target compiler and locations of included files are specified.

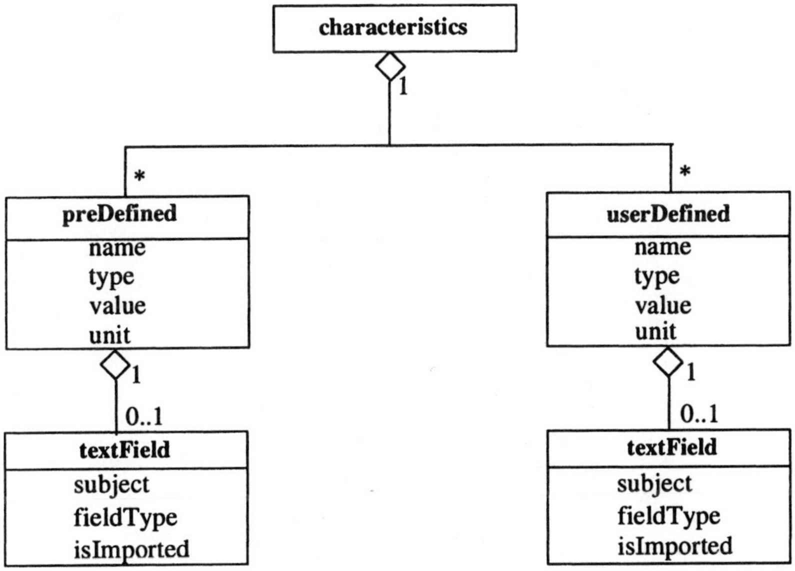

Figure 4-8. Detailed hierarchical structure of the characteristics element.

The following completes the descriptions of the POlif schema elements.

Name	Type	Multiplicity	Description
targetCompiler	Element	1	Expected target compiler
swPackage	Element	0..1	Reference to the software
preDefined	Element	0..*	Pre-defined component characteristics
userDefined	Element	0..*	User-defined component characteristics

Permissible attributes are defined as follows:

Name	Type	Base	Description
name	Enumeration: *PO-dependent*	preDefined	Name of preDefined characteristics
name	CDATA	userDefined	Name of userDefined characteristics
type	Enumeration: *PO-dependent*	preDefined, userDefined	preDefined and userDefined type of characteristics
value	CDATA	preDefined, userDefined	characteristics value
unit	Enumeration: *PO-dependent*	preDefined, userDefined	Unit of characteristics

Chapter 5

UML PROFILE FOR CODESIGN MODELING FRAME (CMF)

5. INTRODUCTION

This chapter describes our framework, that, when taken together with the UML profile for schedulability, performance and time specification, or succinctly the UML real-time profile [29], constitutes the core concept for developing real-time embedded SoC systems in the platform-centric design environment. Our framework, called the Codesign Modeling Framework (CMF), builds upon the real-time foundation provided by the UML real-time profile; it exists as a *supplemental* package, and not as a replacement. The UML real-time profile is, at the time of writing, in the final phase before OMG standardization.

The UML real-time profile offers a facility for modeling and analyzing real-time applications. Such a facility proves adequate for most software development processes. However, in the codesign environment where hardware and software developments often take place simultaneously, the profile becomes less useful—it is less capable of coping with the hardware development, let alone the complexity of the codesign environment where heterogeneous development processes intermingle systematically.

The UML profile for Codesign Modeling Framework, is aimed at providing a solution to such issues. The profile adds to the UML real-time profile the frameworks for modeling exceptions (EMprofile), interrupts (IMprofile), synthesizable HDL (SHDLprofile), as well as an architecture blueprint (ABprofile) that is used as a template to construct the target architecture. The chapter presents details for each profile individually, starting with the utility package (PCUprofile) that provides generic utility extensions for the framework, followed by the EMprofile, IMprofile, SHDLprofile, and ABprofile profiles, respectively.

5.1 Codesign Modeling Framework in Principle

The Codesign Modeling Framework contains a collection of codesign-oriented, real-time profiles whose intent is to enhance the proposed platform-centric SoC design approach. It works in conjunction with the UML real-time profile so as to supplement it with codesign modeling capability. The objectives of the framework detail as follows:

- Permit heterogeneous modelling of hardware and software in the same unified design environment,
- Support modeling and elaborating of an architecture blueprint that results in the target architecture,
- Enable one-to-one mapping of UML to synthesizable hardware description language (HDL),
- Together with the UML real-time profile, provide a standard means for representing LPO tools and components, thus, promoting reuse, and
- Enhance design for reliability by including frameworks for exception and interrupt modeling.

Figure 5.1 depicts the structure of the Codesign Modeling Framework, as well as the relationships among participated packages and actors. In modeling a system for the platform-centric method, the developer derives the target architecture (TargetArchitecture package) from an architecture blueprint (ArchitectureBlueprint package) supplied by the platform provider, i.e. the LPO module provider, and utilizes the derived architecture as the hardware reference for developing software applications (MyModel package). Marked by co-design characteristics, the processes of selecting the target architecture and developing software applications can be performed in sequence, in parallel, or iteratively.

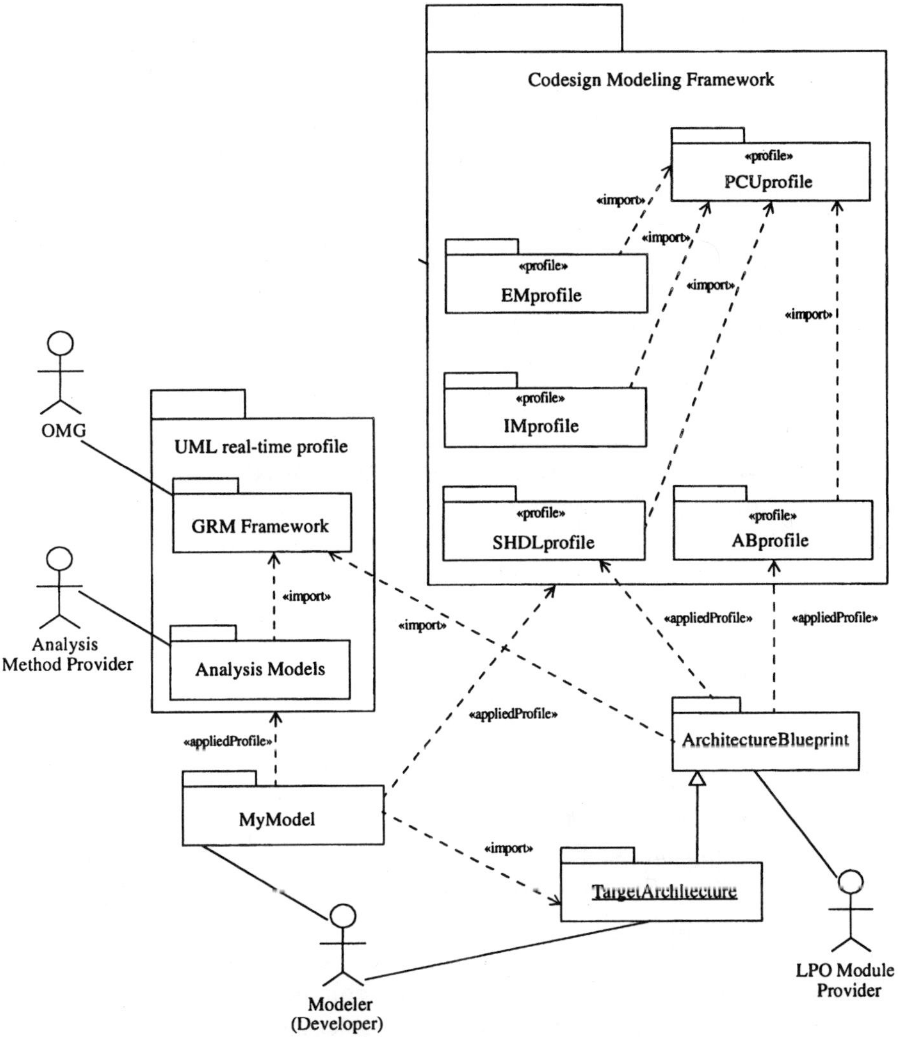

Figure 5-1. Structure of the UML Profile for Codesign Modelin (CMF)g. Also shown are anticipated relationships among participated packages and actors.

In presenting each profile in the Codesign Modeling Framework, this book adopts the two-viewpoint presentation scheme employed in the UML real-time profile specification [29]. The first is the domain viewpoint that "captures, in a generic way, the common structural and behavioral concepts" that characterize each profile. The other is the UML viewpoint, which is "a specification of how the elements of the domain model are realized in UML." The UML viewpoint identifies the required UML extensions, i.e., stereotypes, tagged values and constraints, and groups them in a profile package. Unlike the domain model, however, the UML viewpoint represents the concrete realization rather than the abstract concepts. As such, these extensions do not necessarily map one-for-one with the domain model.

It shall be noted, however, that all of the extensions to be presented are light-weighted, which means they require no fundamental change to UML.

5.2 Platform-Centric Utility (PCUPROFILE)

This section introduces generic utility extensions that serve various purposes to enhance the robustness of the proposed approach. These utility concepts are often disjoint, and are expected to be used by most models and profiles within the platform-centric environment.

5.2.1 Domain Viewpoint

5.2.1.1 Main Function Designation

Employing the Codesign Modeling Framework, the UML models are likely to be mapped into more than one language, e.g. C# for application and synthesizable Verilog for hardware implementation. Automatically determining proper main functions in such models can be complicated for each programming language has its own way of representing the function. For example, C/C++/Java use the keyword main, and Pascal uses program, while such programming languages as Ada/VHDL require explicit designation at link time, and, thus, any procedure/entity can be *main*.

Such complexity can be alleviated by explicitly designating the main function on a proper method in the model, and leaving the language-specific main construct to the code generator. This approach results in the model being more uniform and readable.

5.2.1.2 Link from UML to LPO

In the platform-centric environment, the Codesign Modeling Framework uses and reuses LPO resources, whose existences and availability are identified through the use of POM's services. However, to reduce communication overhead during LPO module import, only the resource models represented in UML should be utilized. To later retrieve relevant information, such models must maintain appropriate links to their parent modules in the LPO.

5.2.1.3 Package Processing Instruction

The developer uses LPO components, i.e. POmm/components, for developing the system by means of UML packages. Not all LPO components, however, have hard characteristics. POmm/components, e.g. software library or legacy code, may need to be modified and/or compiled as part of the development process. In some cases, the UML package may just come raw and the appropriate processing steps can only be determined when the configuration of the hardware platform is known. Therefore, a way for specifying default package processing instruction and/or user-defined instruction is preferable.

5.2.1.4 Code Insertion

Although the Codesign Modeling Framework does allow homomorphic mapping of UML to code upto a certain degree, it has no intent to attain absolute formalism. The efficiency in developing systems of great complexity like today's real-time embedded systems relies also on the principle of design with reuse (DwR) [107, 138] and good analysis tools, both of which do not mesh well with the formalism concept. For analysis, UML already is a great tool. For reuse, it relies on the notion of package.

Code reuse in UML is often tool-dependent. Code is grouped by a UML tool into a library and modeled using UML package. Another means to achieve code reuse in UML is by inserting pieces of code into the model and associating them with the desired methods. Then a code generator can produce full functional source code from the model.

While some better tools support an elaborated code insertion scheme, some do not and only dwell on the mechanism suggested by the UML specification—placing texts of code on a UML note. This simple scheme works fine when local variables can be declared within the method body delimiters like in the languages such as C++ and Java. However, it becomes awkward and more complicated when the declarations have to be done outside of the method body delimiters like in Ada and VHDL. Good tools will still be able to handle it, nonetheless, with more effort. To make code insertion in the Codesign Modeling Framework as general and as uniform to many programming languages as possible, it is recommended that, for each method to be given a piece of code, it designates a declaration area and a body of method area apart from each other, such that a code generator can process the inserted piece of code with little effort and no ambiguity.

5.2.1.5 Non-design Variables

Many times in a course of the development process, the developer will want to use non-design variables for various specific purposes not pertinent to the actual development of the system. An obvious example of non-design variables includes constraint variables that capture non-functional system characteristics such as power dissipation, environmental requirements, and rigid form factors. These variables are predominantly used during the validation process.

Because some profiles and/or stereotypes, e.g. the «SHDLarch», perceive regular design variables as conveying further implicit information, it is required that the tools that will interpret these variables be able to differentiate them from any non-design variable that may be placed in the same context.

5.2.2 UML Viewpoint

In defining UML extensions, i.e. stereotypes, tagged values and/or constraints, for the PCUprofile package, the prefix PCU is always attached to the names to differentiate them from similar or same names in other profiles. This is a standard practice observed in the UML profile for schedulability, performance and time specification, and is exercised here to attain the same clarity effect.

5.2.2.1 Mapping Utility Domain Concepts into UML Equivalents

The *main* function concept maps to the «PCUmain» stereotype attached to an operation (method), and has no tagged value associated with it.

When a UML model (class or object) is derived from a LPO module, it is denoted with the «PCUlpoMember» stereotype. Then, the UML to LPO link concept is reified using either the tagged value PCUuri or PCUid or both. Either one of these tags is enough to identify the corresponding POmm/component in a LPO. Nonetheless, both of them are redundantly furnished for reliability.

The package processing instruction concept maps to the «PCUrun» stereotype on a package. The stereotype has two tagged values, namely PCUrunline, and PCUrunfile, associated with it. The PCUrunline tag specifies a command line to be run against the package. The PCUrunfile tag indicates that processing instructions can be found in the specified file.

The concept of code insertion and reuse maps into the stereotypes «PCUcode», «PCUdeclare», and «PCUcodeBody». A Component can be stereotyped «PCUcode» to indicate that it is a file. The «PCUcode» stereotype defines one tagged value, PCUfileUri that specifies the location of the file. The «PCUdeclare» stereotype permits texts of parameter declaration to be inserted; whereas, «PCUcodeBody» treats the whole textual context as a body of the method.

5.2.2.2 UML Extensions

To minimize the possibility of conflict with other profiles, all extensions in this package are PCU-prefixed. The presentations are of tabular format, as suggested by the UML specification guide [24]. For stereotype tables, the fields include a stereotype name, base class and an associated tagged value. When no tag is defined, it is denoted by --None--. Tag tables include the name, type, multiplicity and domain concept fields.

«PCUattribute»
This stereotype provides a utility to designate design variables, and is particularly useful when used together with the «PCUauxAttr» (see Section 5.2.1.5).

Stereotype	Base Class	Tags
«PCUattribute»	Attribute	--None--

«PCUauxAttr»
This stereotype specifies an auxiliary attribute corresponding to the concept of non-design variables as discussed in Section 5.2.1.5.

Stereotype	Base Class	Tags
«PCUauxAttr»	Attribute	--None--

«PCUcode»
This stereotype provides a file insertion mechanism (see Section 5.2.1.4).

Stereotype	Base Class	Tags
«PCUcode»	Component	PCUfileUri

The following tag is defined:

Tag Name	Tag Type	Multiplicity	Domain Concept
PCUfileUri	String specifying the URI of the file	0..1	See Section 5.2.1.4

«PCUcodeBody»
«PCUdeclare»
The «PCUdeclare» and «PCUcodeBody» represent the declaration, and the body of the code to be inserted into the model, respectively (see Section 5.2.1.4).

Stereotype	Base Class	Tags
«PCUcodeBody»	Note	--None--
«PCUdeclare»	Note	--None--

«PCUconfigList»
This utility stereotype permits configuration attributes to be grouped together separately from their parent class.

Stereotype	Base Class	Tags
«PCUconfigList»	Class	--None--

«PCUlpoMember»
This stereotype specifies an affiliation of a UML model to the LPO. The intent is to furnish a tracing mechanism that will allow relevant information stored in the LPO to be accessible via the UML model itself.

Stereotype	Base Class	Tags
«PCUlpoMember»	Class Object	PCUuri PCUid

The defined tags are:

Tag Name	Tag Type	Multiplicity	Domain Concept
PCUuri	String specifying the URI of the module	0..1	See Section 5.2.1.2
PCUid	String specifying the ID of the module	0..1	See Section 5.2.1.2

«PCUmain»
When adorned on an operation, this stereotype designates the operation to be a main function.

Stereotype	Base Class	Tags
«PCUmain »	Method	--None--

«PCUrun»
The stereotype represents the concept of package processing instruction (see Section 5.2.1.3).

Stereotype	Base Class	Tags
«PCUrun»	Package	PCUrunline PCUrunfile

The tags are defined by:

Tag Name	Tag Type	Multiplicity	Domain Concept
PCUrunline	String specifying the command line	0..1	See Section 5.2.1.3
PCUrunfile	String specifying the file that need to be processed	0..1	See Section 5.2.1.3

The following constraints are defined for this stereotype:

- If the «PCUrun» stereotype is used, at least one of the tags must be used.
- Although it seems redundant, using both tags at the same time is permissible.

«PCUuseConfig»
This stereotype binds a «PCUconfigList» class comprising configuration attributes to their parent class.

Stereotype	Base Class	Tags
«PCUuseConfig»	Dependency	None

5.3 Exception Modeling (EMPROFILE)

In many real-time systems, where dependability is so vital that a failure may be unacceptable; a means to detect errors and faults must be exercised so as to prevent unexpected failures from occurring. Exception provides such a means for system developers, and is a subject of this section. The UML specification [24] does offer a suggestion on how the exception facility should be modeled *per se*. Nonetheless, it comes, not surprisingly, as a general modeling tip for OO programming. This section expands and explores the language horizon, and devises a simple framework that could be used to model exception handling mechanisms in the platform-centric environment.

5.3.1 Domain Viewpoint

Unless indicated otherwise, the discussions in this section and Section 5.4.1, on the Interrupt Domain Viewpoint, are based primarily on a comprehensive survey of real-time systems and programming languages by Burns and Wellings [108].

Exception handling facilities render a means for containing and handling error situations in a programming language. Older programming languages, such as C and RTL/2, have no explicit support for exception handling mechanisms; they rely, instead, on implicit programming techniques such as checking for an unusual return value, and/or programming with a non-local *goto*. Although more recent programming languages, e.g., Ada, Java, often provide explicit support for exception handling facilites, the models adopted by these languages still vary: (1) they may or may not allow an exception to be explicitly represented; (2) an exception may or may not propagate beyond the expected scope of its handler; and (3) parameters may or may not be passed along with the raised exception.

5.3.1.1 Representation of Exceptions

An exception can be detected either by the environment or by an application, and can be raised synchronously or asynchronously. Most mainstream programming languages are sequential in nature, and their exception handling facilities support only synchronous notifications—leaving asynchronous notifications that are mostly employed in concurrent programming to be manually handled by programmers.

Environment-detectable exceptions, on the other hand, often come pre-defined by programming languages, while application-related exceptions normally are user-defined. In a case where an explicit declaration of both exception types is required, they tend to have the same super-type, e.g. a throwable class in Java, or an exception keyword in Ada. Otherwise, a type is only pre-defined for environment-detectable exceptions, and an application can throw any type at all as an exception without pre-declaration. An example of this model is a pre-defined exception class in C++.

From a code generator's point of view, these constructs for representing exceptions, though diverse, can be produced automatically given that the code generator is language-aware, and it is capable of identifying exceptions in the model.

5.3.1.2 Exception Handler Domain

Depending on the context of computation, an exception may be associated with more than one handler. As a result, a domain, or a region of computation, must be assigned to each handler to prevent them from clashing on each other when an exception occurs. A domain is normally associated with a block, subprogram, or a statement. The majority of mainstream real-time programming languages, e.g. Ada, Java, C++, uses a block to specify a domain for a handler.

5.3.1.3 Exception Propagation

When an error event causes an exception to be raised, and there is no handler for it in the enclosing domain, most mainstream real-time programming languages allow such an exception to propagate to the next outer-level enclosing domain. This propagation can continue on until the exception is handled or the program is terminated when no handler for it is found.

5.3.1.4 Parameter Passing

When a programming language permits an exception to be represented as an object, it is normally possible that parameters may be passed along with the exception notification by means of the object attributes.

5.3.1.5 Post-handling Actions

After an exception is raised and handled, the handler may either return the control to its invoker and the computation resumes, or it may terminate the program altogether. The termination model is what most exception facilities adopt, and is the only model considered in this book.

5.3.1.6 Usage Model

The usage models adopted by most mainstream real-time programming languages follow predominantly the throw/try/catch structure. The throw action permits an exception to be raised within an enclosing domain established by the try block. The catch block, then, traps raised exceptions from the associated try block and allows appropriate handlers to be invoked. A specially designated handler is often allowed as a safety measure for the exceptions overlooked by the primary try/catch blocks in order for the program to never fail nondeterministically.

It has been shown by Costello and Truta [109], as well as in the work by Lee [110], that C macros can be used to mimic the throw/try/catch structure, making this popular usage model even more uniform among major programming languages.

5.3.2 UML Viewpoint

In defining UML extensions, i.e. stereotypes, tagged values and/or constraints, for the EMprofile package, the prefix EM is always attached to the names to differentiate them from similar or same names in other profiles.

5.3.2.1 Mapping Exception Domain Concepts into UML Equivalents

To cope with a wide variety of exception modeling characteristics, the profile attempts to model a complete set of relevant information. An exception is generically represented as a signal class adorned with the «EMexception» stereotype, which is a generalization of the standard «signal» stereotype. This stereotype contains no tagged value, and primarily serves to indicate a special requirement for code mapping (see Section 5.3.1.1).

Where the target programming language permits, parameters may be passed by means of class attributes (see Section 5.3.1.4). The exception handler domain concept (see Section 5.3.1.2) maps to the scope imposed by a State Machine diagram containing the try and catch states (see Section 5.3.1.6). The termination model concept (see Section 5.3.1.5) maps to a state transition from the handler state to the *final* state. The exception propagation concept (see Section 5.3.1.3) is viewed as a propagation of the exception signal from the current enclosing State diagram to the next outer-level enclosing State diagram.

As per the usage model (see Section 5.3.1.6), the throw action maps to the «EMthrow» stereotype that defines the EMthrowType tagged value. The EMthrowType tag lists exception types that can be thrown by an «EMthrowMethod» object. The try and catch blocks map to the stereotypes «EMtry», and «EMcatch» in the State Machine diagram, respectively. The handler that is specially designated to trap all other exceptions that are raised without being caught by the «EMcatch» stereotype is represented by a state adorned with the «EMcatchAll» stereotype (see Section 5.3.1.6).

5.3.2.2 UML Extensions

To avoid any possible duplicate and ambiguity, all extensions defined in this profile are prefixed with EM.

«EMbind»
This stereotype binds the exception to its throw and handler class, thus effectively modeling the exception mechanism at a higher-level of abstraction (see Section 5.3.1.6).

Stereotype	Base Class	Tags
«EMbind»	Dependency	--None--

«EMcatch»
This stereotype models the catch structure as discussed in Section 5.3.1.6.

Stereotype	Base Class	Tags
«EMcatch»	SimpleState CompositeState	--None--

«EMcatchAll»
When used after the «EMcatch» block, this stereotype allows the exceptions that are raised but not caught by the catch block to be trapped. In this context, its function resembles the finally and others clauses in Java and Ada, respectively. However, when used alone, it will catches all exceptions, and thus, can be translated to catch(...) in C++.

Stereotype	Base Class	Tags
«EMcatchAll»	SimpleState CompositeState	--None--

«EMexception»
This stereotype models the exception representation (see Section 5.3.1.1).

Stereotype	Base Class	Tags
«EMexception»	Signal	--None--

«EMhandler»
This stereotype models the exception handling concept (see Section 5.3.1.1).

Stereotype	Base Class	Tags
«EMhandler»	Method	--None--

«EMthrowMethod»
This stereotype models the throw statement part of the throw/try/catch structure (see Section 5.3.1.6).

Stereotype	Base Class	Tags
«EMthrowMethod»	Method	EMthrowType

It defines the following tagged value:

Tag Name	Tag Type	Multiplicity	Domain Concept
EMthrowType	TVL List of throwable exception types, for example ('rErr','wErr', 'rwErr')	0..1	Throw statement (see Section 5.3.1.6)

«EMtry»
It represents the try structure concept as discussed in Section 5.3.1.6.

Stereotype	Base Class	Tags
«EMtry»	SimpleState CompositeState	--None--

5.3.2.3 Example

The UML usage model for the exception profile utilizes a State diagram to model the throw/try/catch structure, and attaches it to a method. This method can be nested and called from within another State diagram, resulting in a exception propagation hierarchy.

State Machine diagrams in Figure 5.2 demonstrate the UML usage model of the exception modeling profile. Figure 5.2 (a) shows the UML representation of an exception, along with corresponding Java code excerpted from Burns and Wellings [108]. It is to note in this figure that, even though, the exception possesses no parameter, i.e. class attribute, using and passing parameters along with the exception notification is perfectly doable. The State Machine diagram in Figure 5.2 (b) illustrates the modeling of a throw statement. The try/catch blocks in Figure 5.2 (c) are illustrated as simple states. However, they are applicable to composite states as well. The corresponding Java code for the try/catch states is shown in the figure.

5.4 Interrupt Modeling (IMPROFILE)

Developers of real-time embedded systems so often requires low-level interrupt service programming to implement applications such as device drivers and controllers. The proposed platform-centric SoC design method expects that the developer would be able to mostly avoid such tedious programming by resorting to available hardware-dependent software packages that accompany LPO hardware components. Nonetheless, given the sheer size and diversity of real-time embedded system characteristics, the availability of these packages is far from being a cure-all remedy. There will always exist the times that no suitable packages can be found, and the interrupt programming becomes inevitable, resulting in an increased complexity. The IMprofile package attempts to provide for the developer an interrupt modeling framework that can help ease the complexity incurred by the interrupt programming process, and that is independent of any programming language in particular.

5.4.1 Domain Viewpoint

The actual implementation of interrupt service facilities differs from one programming language to another. However, the underlying requirements for these languages remain the same. They are: device register representation and manipulation, device encapsulation, and interrupt handler. Subsequent discussions are based predominantly on Chapter 15 of the book by Burns and Wellings [108], that details the technical survey on low-level programming.

5.4.1.1 Interrupt Representation and Characteristics

Although many representations of interrupts are possible [108], in principle, they can be viewed simply as a specialized signal. These interrupt signals can be assigned priority levels. They can also be associated with unique IDs that permit the interrupt handlers to take proper actions when interrupt events occur. Certain interrupts in Ada; the clock interrupt for example, are reserved and have no user-defined handlers associated with them. Reserved interrupts are handled through the run-time support system of the language.

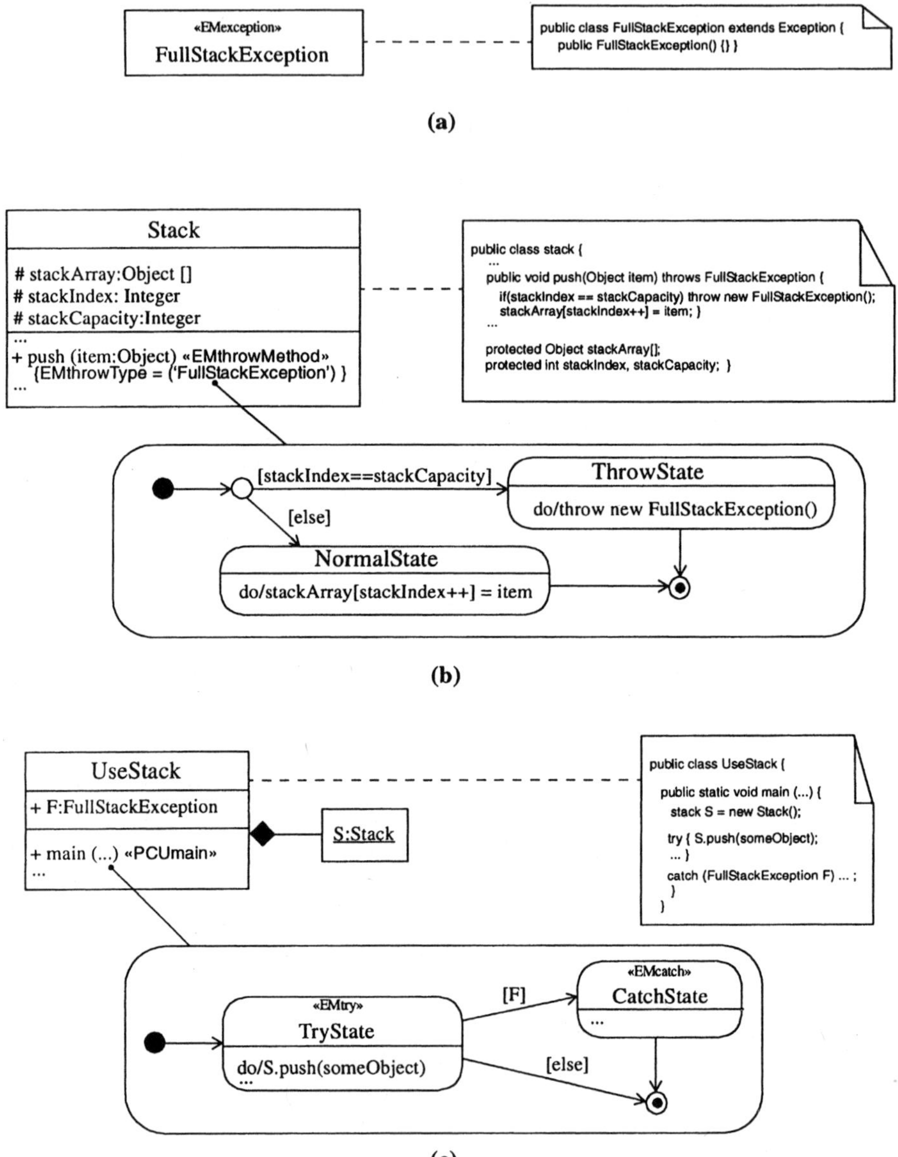

Figure 5-2. Example of UML Exception Modeling Using «EMprofile»

5.4.1.2 Device Register Representation

Each device supported on a hardware platform has as many different types of register as are necessary for its operation [108]. A device register is memory-mapped, and accessible through a memory address. Depending on the hardware platform configurations, a device register may (1) either be oriented most-significant-bit first (descending order, big Endian), or least-significant-bit first (ascending order, little Endian), and (2) be aligned to a specified number of bits, e.g. 8-bit (byte-aligned), 16-bit.

In practice, a device register is divided up into several fields. Each of these fields contains information that is necessary for a correct operation of the device, e.g. control and status data. An accessibility control can be specified such that each individual field can independently set a permission for read-only (r), write-only (w), or both read-write (rw) operations.

To successfully model an interrupt service facility, an expressive way to represent, and manipulate these device registers is a prerequisite. The requirements are: (R1) it shall be able to expressively represent device registers at the bit level, and (R2) it shall include a facility that provides support for bitwise operations. These requirements will be discussed in terms of VHDL as follows.

The requirement R1 involves the support for BIT and BITVECTOR data types—a facility that is extensively supported in most HDL languages. BIT is a scalar type, and has the values '1' and '0', representing logical '1' and '0', respectively. BITVECTOR is then defined as an array of data type BIT. A BITVECTOR can be of a specified size, range, and order—e.g. a descending BIT array of size 8 starting from bit 7 down to 0.

Access to data of type BITVECTOR shall be allowed both on an individual-index basis, or a range-of-index basis. The following bitwise operators shall be supported (requirement R2):

- Logical operators: and or nand nor xor xnor not
- Relational operators: = /= < <= > >=
- Shift operators: sll srl sla sra rol ror
- Concatenating operator: &

When applying logical and relational operators to BITVECTOR operands, the operation is carried out bit-by-bit, and matching position-to-position, until a decisive outcome results. The sll and srl are logical shift left and right. The sla and sra are arithmetic shift left and right, while the rol and ror are left and right rotation, respectively. These operators take the left operand to be a data of type BITVECTOR, and the right operand is an integer indicating a number of positions to shift. The logical shifts fill the vacated bits with logical '0's, while the arithmetic left and right shifts fill the vacated bits with the right-most and the left-most bits, respectively. The sole adding operator takes two operands, either BIT or BITVECTOR, concatenates them, and returns a BITVECTOR as the output.

Assume that all operands are of type BITVECTOR, Table 5.1 demonstrates the use of these operators.

5.4.1.3 Interrupt Handler

An interrupt handler is a software managing routine that executes appropriate actions in response to an interrupt event. Each interrupt signal is bound to a certain handler such that, when an interrupt event occurs, the correct handler becomes active.

5.4.1.4 Device Encapsulation

Interrupt service programming often involves low-level hardware operations that is machine-dependent, and is not portable in general. For a software systems, "it is advisable to encapsulate all the machine-dependent code into units which are clearly identifiable so that separation of portable and non-portable sections are achieved [108]." Examples of such units are classes and packages in Java, protected type facilities in Ada, and a file in C.

5.4.2 UML Viewpoint

In defining UML extensions, i.e. stereotypes, tagged values and/or constraints, for the IMprofile package, the prefix IM is always attached to the names to differentiate them from similar or same names in other profiles.

Table 5-1. Demonstrative user of some bitwise operations

Operations	Results	Operations	Results
not "101"	"010"	"011" xor "101"	"110"
"101" > "111"	FALSE	"010" & "00010"	"01000010"
"10100110" sll 2	"10011000"	"10100110" srl 3	"00010100"
"10100110" sla 2	"10011000"	"10100110" sla 3	"11110100"
"10100110" rol 2	"10011010"	"10100110" ror 3	"11010100"

5.4.2.1 Mapping Interrupt Domain Concepts into UML Equivalents

The interrupt model is attained by stereotyping a Class with the «IMinterrupt» stereotype, where the following tagged values are defined: IMpriority, IMisReserved, and IMid. The interrupt ID maps to IMid, the priority to IMpriority, and a reserved interrupt is specified by the IMisReserved tag (see Section 5.4.1.1).

The interrupt handler concept maps to the «IMhandler» stereotype attached to a method, where it can be bound to an «IMinterrupt» using the «IMbind» stereotype. A class containing the «IMhandler» method represents the device interface block that encapsulates the device-dependent operations together in one place (see Section 5.4.1.4). This Class must be stereotyped with the «IMdeviceIF» stereotype.

5.4.2.2 Mapping Data Type into UML Equivalents

The BIT and BITVECTOR data types map to the IMbit Enumeration, and the «IMbitVector» stereotype, respectively. Their definitions, usages, and constraints are described in detail as follows.

IMbit
It is an enumeration whose permissible values, '0' and '1', represent the logical '0' and '1', respectively.

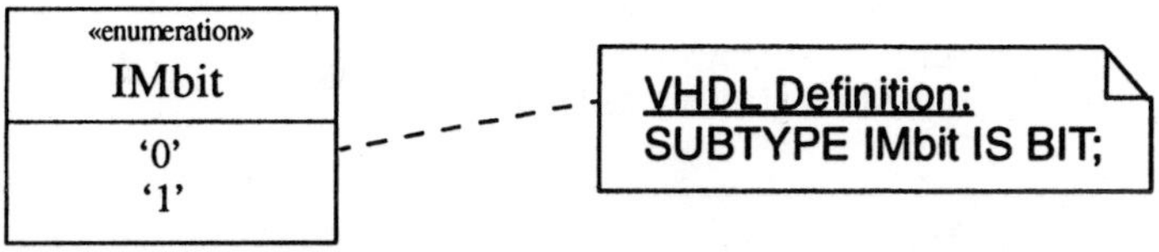

Device Register Representation

To represent device registers in UML, the «IMbitVector» and «IMbitField» stereotypes are used in conjunction (see also Section 5.4.2.3). The «IMbitVector» stereotype is adorned on a Class to designate it as a container of a BITVECTOR data. A BITVECTOR data is represented as an attribute (or attributes) with the «IMbitField» stereotype attached to it. The type expression of the BITVECTOR data follows the format below that satisfactorily expresses size and range of the data (see Section 5.4.1.2):

field_name '[' from_bit ':' to_bit ']' ':' IMbit '[' size ']' «IMbitField»

In addition, when multiple bit fields of the same characteristics are declared in succession, only one «IMbitField» stereotype may be used prior to the first bit field declaration to imply a declaration block. The «IMbitField» block terminates where it encounters another stereotype, or where it reaches the end of the compartment.

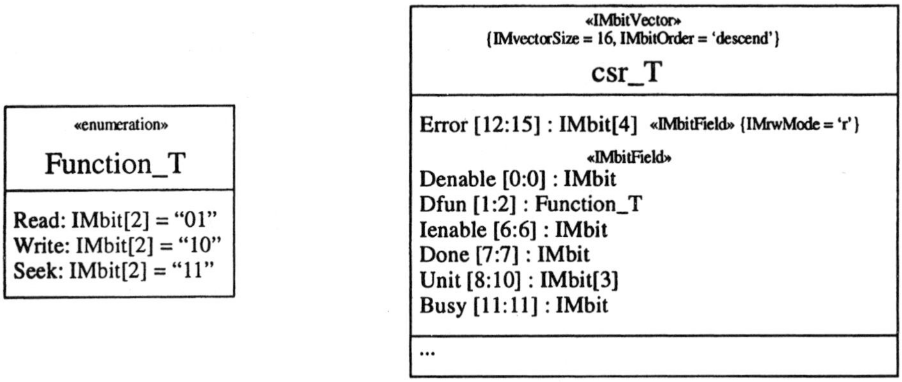

Figure 5-3. Example of the UML representation of a control and status register

A typical control and status register for the computer has the following structure [108], whose equivalent UML representation is shown in Figure 5.3.

```
Bits 15 – 12 : Errors   -- Errors
     11      : Busy     -- Busy
     10 - 8  : Unit     -- Unit select
     7       : Done     -- Done/Ready
     6       : Ienable  -- Interrupt enable
     5 - 3   :          -- Reserved
     2 - 1   : Dfun     -- Device function
     0       : Denable  -- Device enable
```

Also, by using «IMbitVector» and «IMbitField» stereotypes to represent a device register. it implies that:

- If all bit fields can be read from, there must exist a getRegValue() method which returns the register value that results from concatenating all bit fields together. This function is defined in the «IMbitVector» class as:

  + getRegValue: IMbit[IMvectorSize]

- Similarly, if all bit fields can be written to, there must exist a setRegValue() method that takes an IMbit array of size IMvectorSize as the input, and assigns its value to the corresponding bit fields. This function is defined in the «IMbitVector» class as:

  + setRegValue(v: IMbit[IMvectorSize])

- There always exists the getSizeBitfieldName() method that return the length of the bit field indicated by the BitfieldName. This method is defined in the «IMbitVector» class as:

  + getSizeBitfieldName: Integer

Table 5-2. Definition of **IMoppak** operators

Op	**IMoppak Operators**	**Op**	**IMoppak Operators**
and	IMand(v1: IMbit_T, v2: IMbit_T) : IMbit_T	**or**	IMor(v1: IMbit_T, v2: IMbit_T) : IMbit_T
nand	IMnand(v1: IMbit_T, v2: IMbit_T) : IMbit_T	**nor**	IMnor(v1: IMbit_T, v2: IMbit_T) : IMbit_T
xor	IMxor(v1: IMbit_T, v2: IMbit_T) : IMbit_T	**xnor**	IMxnor(v1: IMbit_T, v2: IMbit_T) : IMbit_T
not	IMnot(v: IMbit_T) : IMbit_T	=	IMisEqual(v1: IMbit_T, v2: IMbit_T) : Boolean
/=	IMisNotEqual(v1: IMbit_T, v2: IMbit_T) : Boolean	>	IMisGreater(v1: IMbit_T, v2: IMbit_T) : Boolean
>=	IMisGreaterEqual(v1: IMbit_T, v2: IMbit_T) : Boolean	<	IMisLess(v1: IMbit_T, v2: IMbit_T) : Boolean
<=	IMisLessEqual(v1: IMbit_T, v2: IMbit_T) : Boolean	**sll**	IMsll(v: IMbit_T, k: Integer) : IMbit_T
srl	IMsrl(v: IMbit_T, k: Integer) : IMbit_T	**sla**	IMsla(v: IMbit_T, k: Integer) :IMbit_T
sra	IMsra(v: IMbit_T, k: Integer) : IMbit_T	**rol**	IMrol(v: IMbit_T, k: Integer) : IMbit_T
ror	IMror(v: IMbit_T, k: Integer) : IMbit_T	**&**	IMcat(v1: IMbit_T, v2: IMbit_T) : IMbit_T

- There always exist the setBitfieldName() and getBitfieldName() methods associated with each bit field, unless it is marked as read-only (IMrwMode = 'r'), or write-only (IMrwMode = 'w', see Section 5.4.1.2). In such a case, only the proper method is implemented. For example, assume that the Errors bit field is read-only while the Ienable can be read from or written to, then the following methods are defined:

  + getSizeErrors: Integer

+ getErrors: IMbit[4]
+ getSizeIenable: Integer
+ getIenable: IMbit
+ setIenable(v: IMbit)

5.4.2.3 Mapping Operators into UML Equivalents

In mapping the operators described in Section 5.4.1.2 into UML equivalents, a utility class IMoppak is defined as a grouping mechanism. The operators placed inside the IMoppak class become known globally. Let IMbit_T be a super type of types IMbit and IMbit[]. Then, the operators belonging to the IMoppak class are defined as follows (see Table 5.2).

5.4.2.4 UML Extensions

To avoid any possible duplicate and ambiguity, all extensions defined in this profile are prefixed with IM.

«IMbind»
The stereotype models the explicit binding between an interrupt and its handler (see Section 5.4.1.3). The UML representation shows as an attachment of the «IMbind» stereotype on a dependency between «IMdeviceIF» and «IMinterrupt» classes.

Stereotype	**Base Class**	**Tags**
«IMbind»	Dependency	--None--

«IMbitField»
The stereotype represents an individual bit field within a device register (see Section 5.4.1.2, see also Section 5.4.2.2).

Stereotype	**Base Class**	**Tags**
«IMbitField»	Attribute	IMrwMode

The tag defined for it is:

Tag Name	**Tag Type**	**Multiplicity**	**Domain Concept**
IMrwMode	Enumeration: ('r', 'w', 'rw') Default value: 'rw'	0..1	Accessibility Mode of Register's Bit Field (see Section 5.4.1.2)

«IMbitVector»
This stereotype represents the BITVECTOR data type as discussed in Section 5.4.1.2 (see also Section 5.4.2.2). Just like the BITVECTOR data type, «IMbitVector», together with «IMbitField», are used to model the device register concept in UML.

Stereotype	Base Class	Tags
«IMbitVector»	Class Object	IMaddress IMalignment IMbitOrder IMvectorSize

The definition of each tagged value is presented as follows:

Tag Name	Tag Type	Multiplicity	Domain Concept
IMaddress	Integer	0..1	Address location of the register device (see Section 5.4.1.2)
IMalignment	Integer	0..1	Number of bits in Bit-alignment (see Section 5.4.1.2)
IMbitOrder	Enumeration: ('ascend', 'descend')	0..1	Bit orientation (see Section 5.4.1.2)
IMvectorSize	Integer	1	Number of bits in a device register

«IMdeviceIF»
This stereotype represents the device encapsulation concept as described in Section 5.4.1.4.

Stereotype	Base Class	Tags
« IMdeviceIF »	Class Object	--None--

«IMhandler»
This stereotype represents the device handler concept as discussed in Section 5.4.1.3.

Stereotype	Base Class	Tags
« IMhandler»	Method	--None--

The following constraint is defined for this stereotype:

- A «IMhandler» method must reside in the «IMdeviceIF» class.

«IMinterrupt»
This stereotype models the interrupt concept (see Section 5.4.1.1).

Stereotype	Base Class	Tags
«IMinterrupt»	Class Object	Imid ImisReserved IMpriority

The following tagged values are defined:

Tag Name	Tag Type	Multiplicity	Domain Concept
IMid	String	0..1	Interrupt ID (see Section 5.4.1.1)
IMisReserved	Boolean	0..1	Reserved interrupt (see Section 5.4.1.1)
IMpriority	Integer	0..1	Interrupt priority (see Section 5.4.1.1)

5.4.2.5 Usage Model Framework

Figure 5.4 delineates a possible usage model framework for the Interrupt Modeling Profile (IMprofile) package. In the figure, InterruptInterface and AnInterrupt represents a design template that the developer can quickly generalize for the modeling of device-specific operations. This template, which specifies relevant tagged values, and the device-specific run-time library are usually supplied by the LPO component provider.

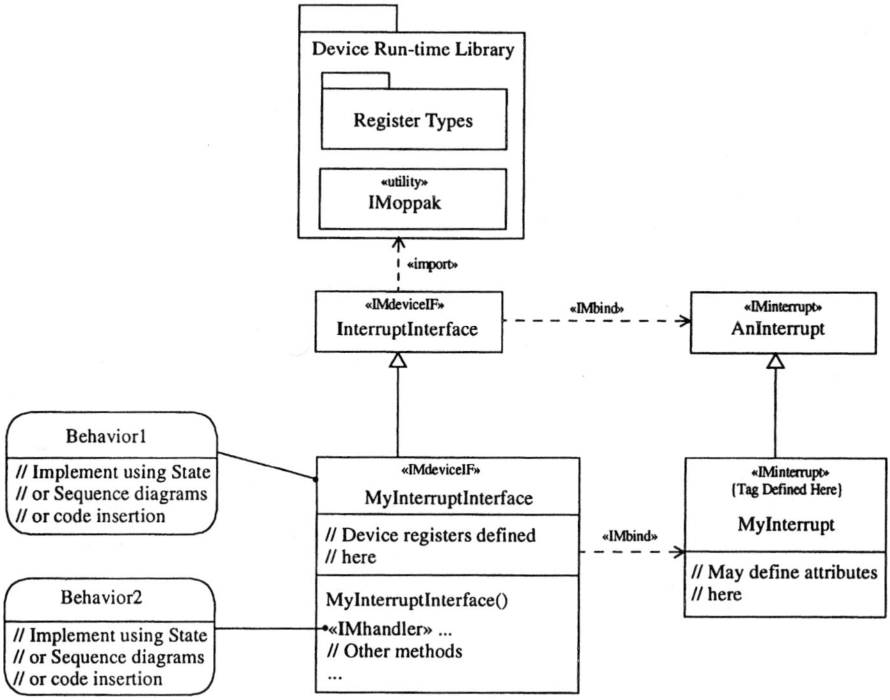

Figure 5-4. Usage model framework for the Interrupt Modeling profile

The device encapsulation block, MyInterruptInterface, provides an interface facility between the software system and the device. Its behavior can be described using State or Sequence diagrams, or code insertion (Behavior1). Within its region of computation, defined are instances of device registers, as well as the interrupt handler whose functionality is described by the Behavior2 module. The «IMbind» stereotype on a dependency between MyInterruptInterface and MyInterrupt classes practically binds the handler method in MyInterruptInterface to the interrupt represented by the class MyInterrupt. Using this framework, a simple homomorphic mapping from the interrupt model to code becomes possible as shown in Table 5.3.

5.5 Synthesizable HDL Modeling (SHDLprofile)

Thus far, this chapter has established a bridging mechanism (Section 5.2) that permits information from the LPO to be retrieved and used in the UML context that encompasses system modeling, analyzing, and implementing. Thereafter Sections 5.3 and 5.4, introduce the Exception and Interrupt

Modeling profiles that, together with the upcoming UML real-time profile, provide a comprehensive support for the development of platform-based real-time applications. This section embarks on the issues of hardware implementation, and presents the Synthesizable HDL Modeling (SHDLprofile) package that enables the use of UML to descriptively model hardware for synthesis. With the SHDLprofile package, both hardware and software development can take place in one unified environment, rendering the development task less complicated.

5.5.1 Domain Viewpoint

In the Electronic Design and Automation (EDA) realm, Hardware Description Languages (HDL) are used to describe hardware functionalities, verify functional correctness, and synthesize the code for subsequent fabrications. The synthesis capability of today's mainstream HDL languages, i.e., Verilog and VHDL, has advanced so much from a few years back that it becomes very commonplace and almost indispensable in the development of any hardware system. In implementing the SHDLprofile package, only the synthesizable subset of HDL syntax and semantics is considered.

Table 5-3. Interrupt model to code mapping

Model Element	Real-Time Java [108, 111]	Ada [108]
«IMdeviceIF»	Class	Protected procedure. If the model element is a template class, a parameter can be passed to the constructor, and thus, the protected procedure.
«IMpriority»	Class *PriorityParameters*	pragma Interrupt_Priority(IMpriority)
«IMhandler»	Class *AsynchEventHandler*	Procedure
«IMbind»	MyInterrupt.addHandler(IMhandler) MyInterrupt.bindTo(IMid)	pragma Attach_Handler(IMhandler, IMid)
«IMinterrupt»	AsynchEvent MyInterrupt	Ignored. However, it signifies the requirements for various Ada interrupt packages. Moreover, tagged values are used.
Registers	RawMemoryAccess	var
Methods in IMdeviceIF	methods	entry procedures
Behavior1	Either ignored, or becomes a controller method in parent class.	Initialization procedure for the protected procedure.
Behavior2	method body	procedure body

HDL languages differ from traditional software-oriented programming languages in many aspects. Of all the discrepancies, it likely is the concepts of signals, time, and concurrency that renders the dissimilitude quite notable. Such concepts require specialized data types, data values and language constructs that make the UML modeling of HDL appears extraneously awkward, let alone the fact that UML is fully object-oriented, while most, if not all, HDL languages are not. Even with the current real-time-oriented UML profile for Schedulability, Performance and Time Specification, UML is still deemed inadequately expressive for the purpose of describing hardware.

The Synthesizable HDL profile, tailored specifically for the proposed approach, is aimed to augment UML with the capability to expressively describe hardware by means of UML models, such that unambiguous UML-to-HDL mapping can be realized to aid the design and implementation of hardware systems. The profile, however, does not attempt to strictly formalize UML for formalism tends to hinder the analysis capability which is so essential for the proposed approach. Instead, the profile relaxes the strict formalism to allow code insertion where appropriate to expedite the development process.

HDL synthesizability is considerably dependent on the interpretation by synthesis tools; there really is no perennial guarantee for interoperability. As such, this section relies on several sources, books, tutorials, papers, and manuals alike, to define the common working subset of synthesizable HDL that is likely to yield the same synthesis results. As per VHDL, the IEEE 1076.6 standard for VHDL Register Transfer Level (RTL) Synthesis [112] serves as the ultimate reference. On the other hand, the interoperability standard for synthesizable Verilog is still work in progress (see http://www.vhdl.org/vi/vlog-synth/). Every effort is made to ensure the best possible interoperability of synthesizable Verilog in this book. The basis of subsequent discussions related to VHDL and Verilog comes predominantly from [113], [114], [115].

Little known use of UML modeling for hardware design has been documented. Much attention has been directed to formalizing UML object and dynamic models so as to automate the code generation process. Björklund and Lilius [116] have demonstrated that automatic generation of optimized synthesizable VHDL code from UML State diagrams could be achieved. In McUmber and Cheng [117], a method to formalize UML object and dynamic models that allows in-model VHDL simulation is reported. Nonetheless, none of these researches provides UML facilities comprehensive enough to tackle the issue of UML modeling for hardware design, which shall (1) allow hardware systems to be described structurally and behaviorally for proper analysis, and (2) include comprehensive facilities for modeling specialized HDL data types, data objects, operations, and language constructs. Rather, the profile bears some conceptual resemblance to the executable UML concept described in Mellor [118], but is more complete as an aid for hardware design and is specifically customized to work well in the platform-centric design environment. No concrete work regarding the UML modeling of Verilog HDL has been found. Such passive research activities in this field can probably be attributed to the almost non-OO nature of Verilog that may prove less attractive compared to the object-based VHDL. It is expected, though, that research activities will increase as Verilog-2005 rolls out [119] and UML 2.0 becomes mature.

5.5.1.1 HDL Design Entities

The VHDL design entities are more modular and more complicated per se when compared with the Verilog counterparts. As such, the discussion herein will proceed solely in terms of VHDL.

VHDL specifies five different design units: entity, architecture, configuration, package and package body, as illustrated in Figure 5.5 [113]. An entity/architecture pair, known collectively as a design entity, is the main construct for describing a hardware component; it is analogous to Verilog's module. The entity provides the port information of a particular design entity, while the architecture provides the functional body description. The configuration functions as a binding mechanism that associates an entity to a particular architecture. Finally, the package and package body hold common design data that can be used by a design entity. Their Verilog equivalent is the include construct.

5.5.1.2 Data Types, Data Objects, and Operations

Unlike the IMprofile package that explicitly defines bit-oriented data types, and operations (see Sections 5.4.2.2 and 5.4.2.3) in order to mitigate the difficulty in modeling interrupt service routines, the SHDLprofile package, instead, adopts the data types, data objects, and operations of the target language.

Whereas VHDL defines a number of data types, Verilog only defines one and it never has to be explicitly specified in any declarative statement. A data object is a mechanism adopted by both VHDL and Verilog to pass data from one point to another. Examples of a data object in VHDL are signal and variable, while examples in Verilog are wire, reg (register), and parameter. A code generator is expected to be able to understand, syntactically and semantically, the language-specific declarations and descriptions in the model.

5.5.1.3 Code Structure

The HDL code structure typically comprises declarative statements, sequential statements, and concurrent statements.

Like most other programming languages, the declarative statements declare objects for use in other parts of the design unit. In VHDL, these statements are always located before the begin clause in a package body, architecture, process, procedure, or function statement. Verilog, on the other hand, has no dedicated declarative region.

Statements after the begin clause in the process statement (VHDL), and inside the always statement (Verilog) are sequential. So are those in the procedure (VHDL), task (Verilog) and function (VHDL and Verilog) statement.

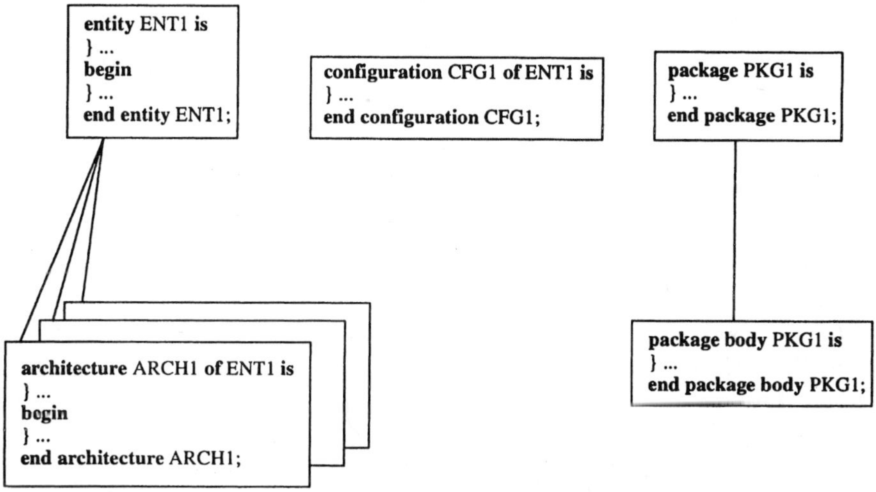

Figure 5-5. VHDL design units

As per concurrent statements, the supported set for VHDL/Verilog encompasses process/always statements, signal/continuous assignments, and VHDL procedure calls. In order to execute concurrently, VHDL signal assignments and procedure calls must not be nested inside a process statement.

5.5.1.4 Behavioral Description

Sequence of HDL statements, that may execute sequentially or concurrently within a design unit, constitute a behavior of that design unit. In VHDL, the following behavioral constructs are synthesizable: process, wait on, wait until, procedure, function, if clause, case clause, for loop, generate, next, and exit. In Verilog, the list includes: always, @( ), task, function, if clause, case clause, and for loop. The collaboration among design units is achieved by means of port and signal (or reg in Verilog).

It is often the case in VHDL that only the wait until statement is permissible for synthesis. In addition, the process and procedure body can only contain at most one wait statement, usually as the first statement in the sequence. In the case of the task statement in Verilog, it can contain no wait equivalent, i.e. @( ), at all. Such a diverse usage of wait statements can result in a modeling complication. As such, it is often a good practice to exclude all synchronous wait statements from the body of subprograms and/or process/always statement, and use VHDL's sensitivity list or Verilog's @( ) statement, instead.

A concurrent statement, i.e. the process statement, and the statement in the architecture body that do not belong to any process, never terminates. Once it executes to its entirety, it starts over from the beginning. This concept is to be referred to as an endless continuation.

A generate statement in VHDL furnishes a mechanism to render a description of regular concurrent statements compact. A for-generate can be used to replicate such statements a predetermined number of times; whereas, an if-generate stipulates the replication of such statements by means of conditional statements. The generate statements are expanded back to the normal, explicit descriptions during compilation time.

Parametrized design is also possible in both VHDL and Verilog, making use of the generic clause, and parameter overloading, respectively. Reuse is attained through the VHDL package design unit, and the Verilog include clause.

Coding style is also significant since the synthesis tool often infers from it to make synthesis-related decisions. For example, using an if statement can infer a priority encoder to be synthesized, while a case statement inferring a multiplexer, or a mux.

5.5.2 UML Viewpoint

In defining UML extensions, i.e. stereotypes, tagged values and/or constraints, for the SHDLprofile package, the prefix SHDL is always attached to the names to differentiate them from similar or same names in other profiles.

Due to the object-based nature of VHDL, it is more straightforward to present the mapping using VHDL concepts and constructs. Then, where appropriate, related information on the corresponding Verilog counterparts ensue. The presentation is organized by first considering in Section 5.5.2.1

the mapping of the design entity collaboration concept, followed by the mapping of the generic HDL structure and behavioral constructs in Sections 5.5.2.2 and 5.5.2.3, respectively. Thereafter, Section 5.5.2.4 details the UML extensions for the SHDLprofile package. The presentation concludes in Section 5.5.2.5 with a demonstrative example on using this profile.

5.5.2.1 Mapping Design Entity Collaboration Mechanisms into UML Equivalents

The collaboration among design entities involves the port and signal concepts—how to define and represent them in UML, and how ports from different design units are bound together to establish a communication channel by means of the signal data object (reg in Verilog).

A port declaration maps to an attribute declaration in an entity class, with one of the «SHDLin», «SHDLout», «SHDLinout» stereotypes attached to the attribute to indicate the port direction. Each port takes on one of the user-defined types that specifies the HDL-dependent data type and data object type. These types are described in the «SHDLtypedef» tag section. Though not explicitly specified, syntactically all VHDL ports are data objects of type signal. Verilog only defines one base data type, which never has to be declared (see Section 5.5.1.2). In Verilog, I/O ports are viewed as wire. Verilog inputs should be declared a wire, while the outputs can either be declared a wire or a reg.

As an example, an AND2 ports may be represented as follows:

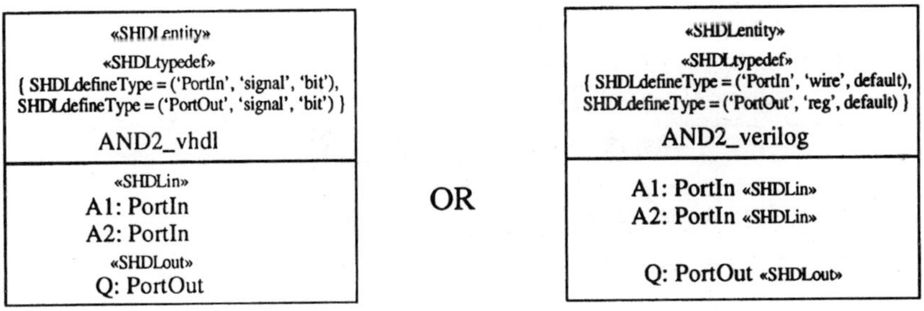

5.5.2.2 Mapping Generic HDL Structures into UML Equivalents

The design entity concept maps to the «SHDLmodule» stereotype that embellishes an abstract class (see Section 5.5.1.1). This stereotype represents a synthesizable HDL domain, and can be iteratively nested. Its presence denotes the existence of the entity/architecture pairs, or other «SHDLmodule» classes. The stereotype defines no tagged value.

The entity and architecture concepts map into a class with the «SHDLentity» and «SHDLarch» stereotype, respectively. In Verilog, the «SHDLentity» stereotype represents the header area of the module, while, the «SHDLarch» stereotype represents the module body. The «SHDLentity» class shall contain no method definition. In fact, it must contain only port definitions that define the design entity interface. The VHDL configuration concept maps to the «SHDLbind» stereotype that binds a pair of «SHDLentity»/«SHDLarch» classes together.

The VHDL generic clause (parameter overloading in Verilog) maps to a «SHDLentity» template class. Parameters in the template window represent the generic parameters in VHDL.

As opposed to the «SHDLentity» that specifies the communication interface for the «SHDLmodule», the «SHDLarch» describes the behavior for it. The VHDL architecture name maps to the name of the «SHDLarch», while architecture-scoped signal declarations map to attribute declarations that are stereotyped by «SHDLdataObject».

Within the architecture body, instantiation of other entities is possible. In VHDL, such a process involves (1) declaring the design entity to instantiate, and (2) instantiating the design entity, and binding the desirable ports and signals together via the port map statement. The design entity concept maps to a user-defined data object, and thus, could be declared in the attribute list compartment. The actual instantiation of the design entity maps to a composition relationship from the «SHDLarch» class to the «SHDLentity» object. port map statements map to port value assignments in the «SHDLentity» object.

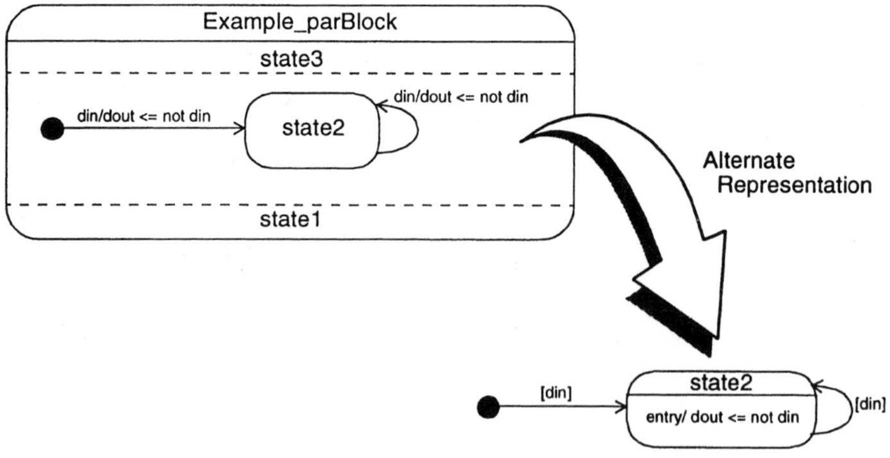

Figure 5-6. Concurrent state representation of **dout <= not din**

The process (Verilog's always) statement maps to an object stereotyped with «SHDLprocess». This object can be implemented by State Machine diagrams, or by inserting source code for it. The relationship between the «SHDLarch» class and the «SHDLprocess» object is defined by a composition association (see Figure 5.7).

The concurrent (Verilog's assign) statements that reside outside the process block in the architecture body, and that do not involve entity instantiations. map to a container object stereotyped with the «SHDLparBlock» stereotype. By treating these statements as having a container object associated with them, they can be annotated with QoS properties and analyzed just like any other objects, if need be. Then there are two ways to associate the unmapped concurrent statements with the «SHDLparBlock» object. The first method is to use code insertion defined in Section 5.2.2.2, i.e. «PCUdeclare», and «PCUcodeBody». Secondly, concurrent State Machine diagrams can be employed. As an example, consider the VHDL code that negates the values of din and assigns the new value to a signal dout:

dout <= not din;   *-- concurrent statement*

The State Machine diagram for this expression is portrayed in Figure 5.6. It has no final state and is perennially active. Every time an event occurs on din, the above statement is evaluated and dout has a new value assigned.

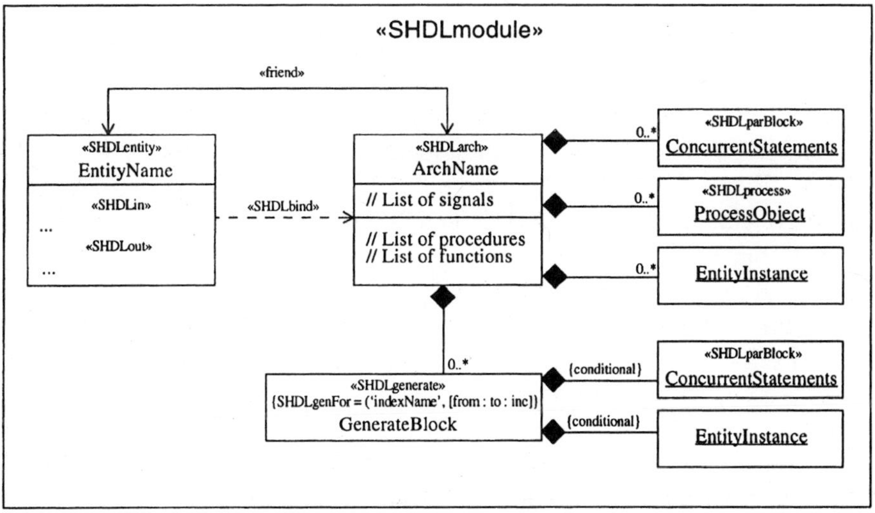

Figure 5-7. Summary of the relationships among entities in the «SHDLmodule»

A generate statement maps to a class being stereotyped with «SHDLgenerate» (see Section 5.5.1.4). The usage model for the «SHDLgenerate» stereotype is portrayed in Figure 5.7. As seen in the figure, an «SHDLgenerate» class may contain one or more instances of either the «SHDLparBlock» or «SHDLentity» object. The for statement is inferred by the SHDLgenFor tag owned by the «SHDLgenerate» stereotype. Conditional statements for the if-generate map to constraints on the Composition.

Procedure and function declarations map to method declarations in the method compartment of the «SHDLarch» class. All methods must be visible publicly (+). A method that has no return value infers a procedure (VHDL) or a task (Verilog), whereas a method that has a return value infers a function (VHDL and Verilog). Procedure and/or function parameters must be of type SHDLdataType. Similar to the concurrent statement, its implementation entails the use of code insertion (see Section 5.2.2.2), and/or State Machine diagrams (see Section 5.5.2.3 for details). Figure 5.7 summarizes the relationships among relevant entities within the «SHDLentity»/«SHDLarch» pair.

The package and package body, where global variables and functional facilities are defined, maps to a class stereotyped with «SHDLpackage». The stereotype has no tagged value. Then the global variables map to attributes in the attribute list compartment; whereas, procedures and functions map to methods in the method compartment. See the discussion above for details about the method inferences of procedures and functions.

5.5.2.3 Mapping Synthesizable HDL Behaviors into UML Equivalents

State Machine diagrams are the UML model of choice for representing behavioral HDL statements. To implement the «SHDLmodule», sequential state models can be attached to the «SHDLprocess» object, and an entity instance. A concurrent state model can be associated with the concurrent statement object («SHDLparBlock»).

The endless continuation concept (see Section 5.5.1.4) maps to a self-iterating state, where there is no final state, and the state always transitions back to itself—pending on the same guard condition as described by the event guard on the transition from the start state to the first state. The state model depicted in Figure 5.6 is one such example.

The process statement maps to a self-iterating composite state as depicted below. Due to its endless continuation characteristic, the composite state contains no final state, and the last sub-state emanates a transition out of the boundary of the composite state before looping back with the same event guard as the one modeled from the sensitivity list.

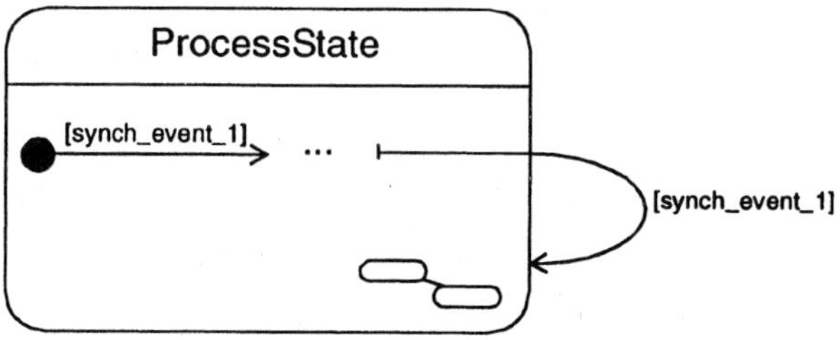

In VHDL, it is required for this profile that if a wait statement is to be utilized in a process and/or procedure, it must be the very first statement and the only wait statement in the process/procedure body. It is recommended that a sensitivity list be used with the process statement, in place of the wait statement. Verilog's task, and VHDL/Verilog's function do not support the synchronization concept.

Then, the wait statement in process/procedure, and/or the sensitivity list concept in the process statement map to the event guard on the transition from the start state to the first state, and the synchronization signals map to the synch_event conditions, as shown below.

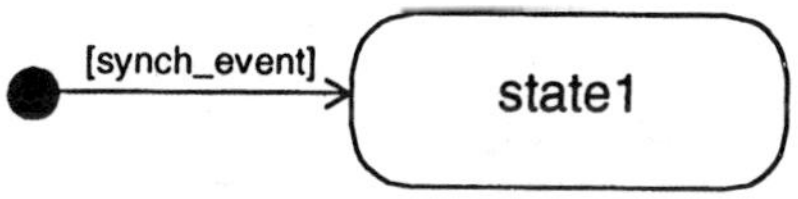

The following illustrate the mapping of HDL behavioral constructs to the UML equivalents. These constructs include: transition on wait conditions, if clause, case statements, for loop, VHDL's next and exit statements, and while loop.

If clause

Order control over the if sequence statements sometimes is significant, and can be modeled as shown below using a dynamic choice path.

```
-- If statements
--
st1;
if g1 then
  st2;
else if g2 then
  st3;
else
  st4;
end if;
st5;
```

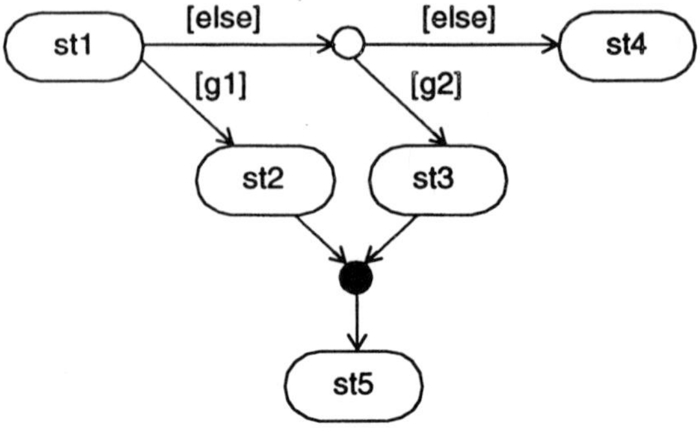

case statement

The standard UML modeling of a case statement is in better agreement than the if counterpart. Generally, the case statement is modeled as illustrated below.

```
-- VHDL
-- st1
case A is
 when ... => action1;  -- st2
 when ... => action2;  -- st3
 others   => action3;  -- st4
end case;


// Verilog
// st1
case(A)
  ... : action1;        // st2
  ... : action2;        // st3
  default: action3;  // st4
endcase;
```

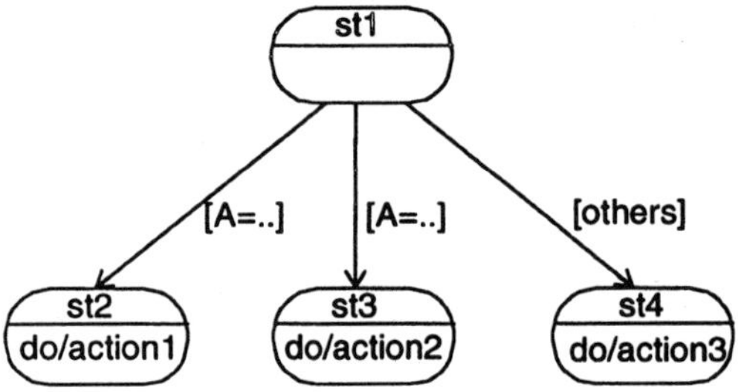

Transition on wait conditions

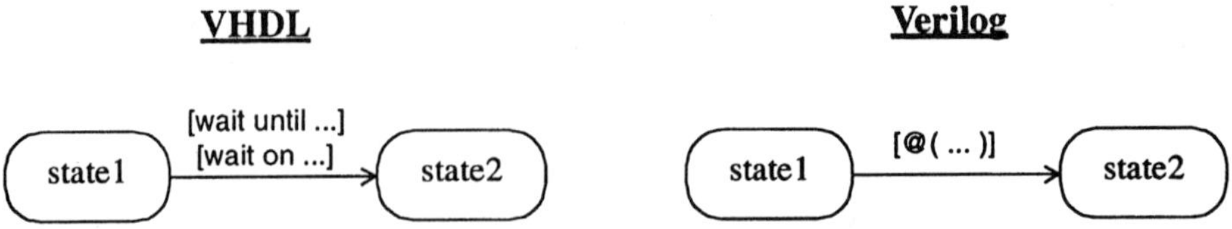

for loop

The for loop model presented here follows that of Rational Rose™ [102] very closely.

```
-- VHDL
--
for i in 0 to 2 loop
  action;
end loop;

// Verilog
//
for(i=0; i<=2; i=i+1)
 begin
   action;
 end
```

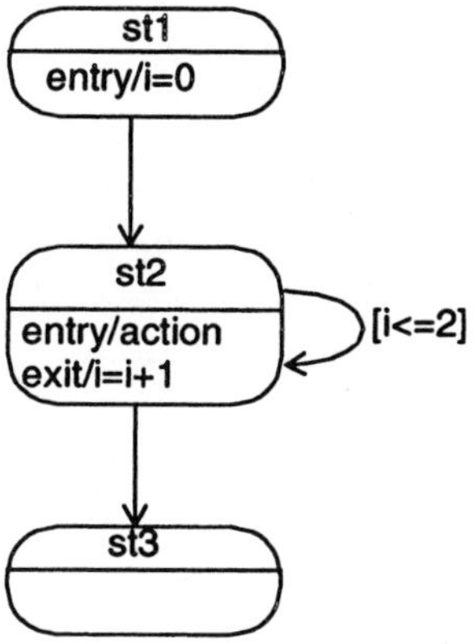

while loop

The while loop, due to its unrestrained nature, is supported by only few synthesis tools (yet, with certain set of rules to follow). However, it merits to be included here for completeness.

```
-- VHDL
--
while B /= 0 loop
  action1;
  action2;
end loop;
```

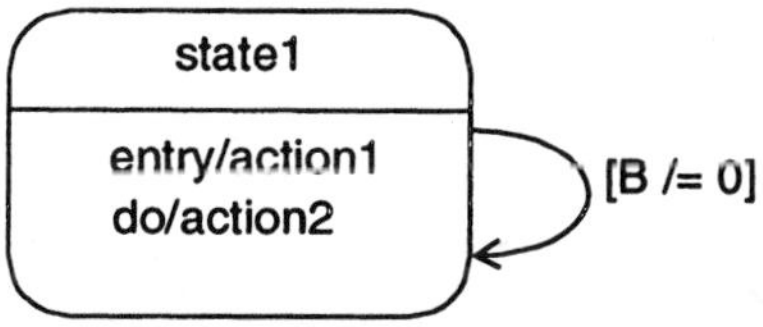

VHDL's next and exit statement

The next statement in a loop maps to a state with only one action: exit/increment loop index. Similarly, the exit statement maps to a state with the action: exit/exit.

next Statement

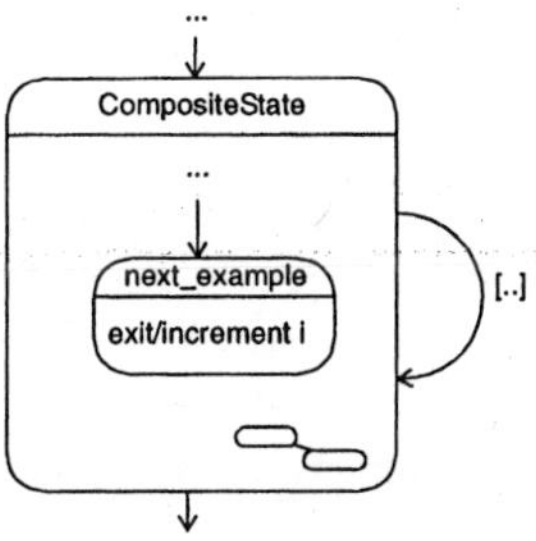

exit Statement

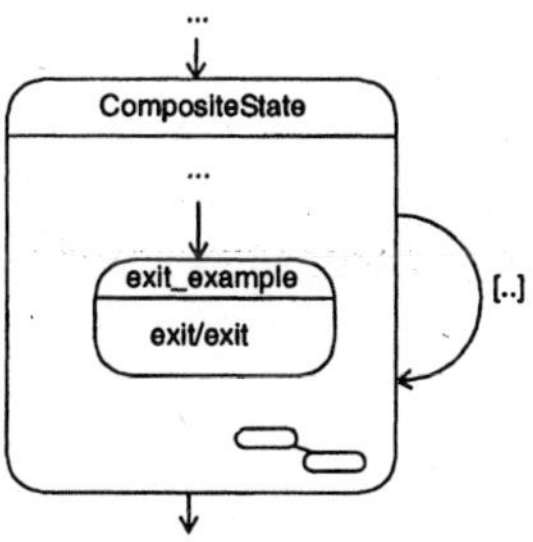

5.5.2.4 UML Extensions

To avoid any possible duplicate and ambiguity, all extensions defined in this profile are prefixed with SHDL.

Port Direction («SHDLin», «SHDLout», «SHDLinout»)
These stereotypes help specify port directions for attributes of the «SHDLentity» (see Section 5.5.2.1).

Stereotype	Base Class	Tags
«SHDLin»	Attribute	--None--
«SHDLout»	Attribute	--None--
«SHDLinout»	Attribute	--None--

The following constraint is defined for these stereotypes:

- Holder of the Attribute base class must be stereotyped «SHDLentity».

«SHDLarch»
This stereotype represents the architecture concept of the entity/architecture pair presented in Sections 5.5.1.1 and 5.5.2.2.

Stereotype	Base Class	Tags
«SHDLarch»	Class	--None--

SHDLattrType
The SHDLattrType allows a tuple of string values that specifies the HDL language-dependent data type and data object type to be mapped to a user-defined attribute type. Its value is described using the Tag Value Language (TVL) as defined in the UML real-time profile specification [29]. To represent the syntax, this book follows the standard BNF notational conventions, where:

- A string between double quotes (") represents a literal,
- A token in angular brackets (< >) is a non-terminal,
- A token enclosed in square brackets ([<element>]) implies an optional element of an expression,
- A token followed by an asterisk (<element>*) implies an open-ended number of repetitions of that element,
- A vertical bar indicates a choice of substitutions.

The TVL uses parentheses to identify arrays, commas to separate elements of arrays, and single quotes for string literals. The SHDLattrType is defined as follows:

```
<shdlAttrTypeStr> ::= (<attrTypeName>,<dataObjStr>,<dataTypeStr>)
<attrTypeName> ::= <String>
<dataObjStr> ::= <String> | "default"
```

```
<dataTypeStr> := <String> | “default”
```

«SHDLbind»
This stereotype represents the binding of the «SHDLentity» and «SHDLarch» classes (see Section 5.5.2.2).

Stereotype	Base Class	Tags
«SHDLbind»	Dependency	--None--

«SHDLentity»
The stereotype represents the entity concept as described in Sections 5.5.1.1 and 5.5.2.2.

Stereotype	Base Class	Tags
«SHDLentity»	Class Object	--None--

SHDLforInfoType
Similar to the SHDLattrType, the SHDLforInfoType is described using the TVL language. It permits necessary information for executing a for-loop, i.e. index parameter name, range, and incrementation step, to be captured through the SHDLgenFor tag (see «SHDLgenerate» below).

The SHDLforInfoType is defined as follows:

```
<shdlForInfoTypeStr> ::= ( <indexNameStr> , <indexRange> )
<indexNameStr> ::= <String>
<indexRange> ::= “[“ <from> “:” <to> “:” <incStep> “]”
<from> ::= <Integer>
<to> ::= <Integer>
<incStep> ::= <Integer>
```

«SHDLgenerate»
This stereotype represents the modeling of VHDL’s generate block (see Section 5.5.2.2).

Stereotype	Base Class	Tags
«SHDLgenerate»	Class	SHDLgenFor

The stereotype defines one tagged value, which is:

Tag Name	Tag Type	Multiplicity	Domain Concept
SHDLgenFor	SHDLforInfoType	1	generate Statement, see Section 5.5.2.2

«SHDLmodule»

This stereotype represents the design entity concept as described in Sections 5.5.1.1 and 5.5.2.2.

Stereotype	Base Class	Tags
«SHDLmodule»	Class	--None--

«SHDLparBlock»
This stereotype represents the grouping of concurrent statements that exist within the architecture body, but outside any process (see Section 5.5.2.2).

Stereotype	Base Class	Tags
«SHDLparBlock»	Object	--None--

«SHDLprocess»
This stereotype represents the existence of a process in the architecture body (see Sections 5.5.2.2 and 5.5.2.3).

Stereotype	Base Class	Tags
«SHDLprocess»	Object	SHDLsensitive SHDLsensitive_pos SHDLsensitive_neg

It defines the following tagged values:

Tag Name	Tag Type	Multiplicity	Domain Concept
SHDLsensitive	TVL List of signals that a process is sensitive to changes, e.g. ('clear', 'reset')	0..1	Sensitivity List (see Section 5.5.2.2 and 5.5.2.3)
SHDLsensitive_pos	TVL List of signals whose positive edge a process is sensitive to	0..1	Sensitivity List (see Section 5.5.2.2 and 5.5.2.3)
SHDLsensitive_neg	TVL List of signals whose negative edge a process is sensitive to	0..1	Sensitivity List (see Section 5.5.2.2 and 5.5.2.3)

«SHDLtypedef»
This utility stereotype enables the HDL-specific data type and data object type associated with a particular data value to be captured using a user-

defined attribute type (see Section 5.5.2.1). The «SHDLtypedef» defines one tag, namely, SHDLdefineType.

Stereotype	Base Class	Tags
«SHDLtypedef»	Class Object Note	SHDLdefineType

The SHDLdefineType tag is defined as follows:

Tag Name	Tag Type	Multiplicity	Domain Concept
SHDLdefineType	SHDLattrType	1..*	HDL-specific data type and data object type, see Section 5.5.2.1

5.5.2.5 Example Usage

To demonstrate the usage model for the SHDLprofile package, specifically the framework depicted in Figure 5.7, an implementation of a six-bit-add-two-bit adder modified from Smith [113] is presented.

The example uses the structural style of hardware modeling to eventually realize a six bit add-two-bit adder. A one-bit half adder is first described that is utilized later to construct a one-bit full adder. Then by properly connecting full adder instances together, the example is able to attain the structural description of a six-bit-add-two-bit adder as desired. Figure 5.8 portrays the UML models of the half adder targeted for (a) VHDL and (b) Verilog; different implementation styles are used for demonstrative purposes.

Although not explicitly displayed in the figure, properly package importation shall be strictly exercised. Figure 5.9 depicts the implementation model of a full adder, using the half adder entities in Figure 5.8. Figure 5.10 shows how the desired six-bit-add-two-bit adder is implemented for the targeted VHDL. Because of its regular structure, the design takes advantage of the generate statement. The model for Verilog should be fairly easy to acquire from the VHDL model, with only one discrepancy: All six FA instances must be explicitly instantiated in Verilog for it does not have a generate equivalent. Specific tools may choose to support the generate model for Verilog; it is purely tool-dependent.

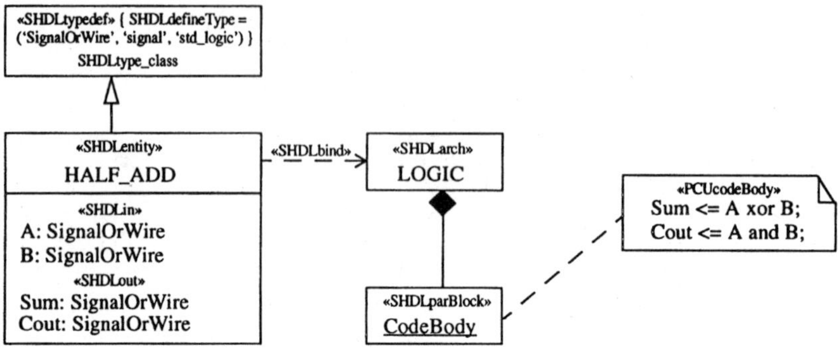

(a) VHDL Model

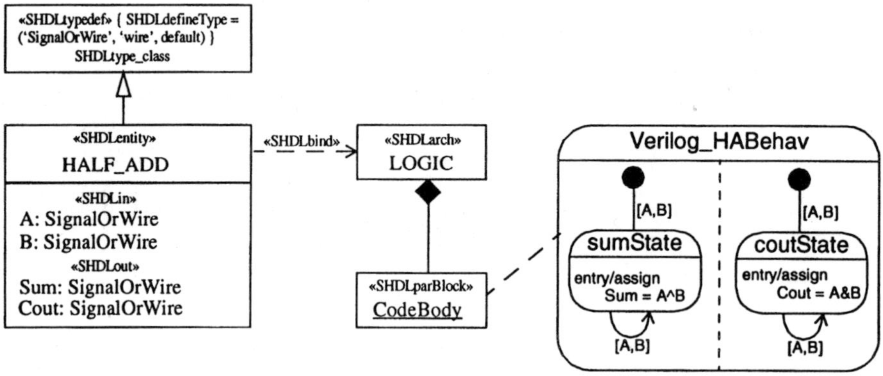

(b) Verilog Model

```
-- VHDL
-- Library usage not shown
--
entity HALF_ADD is
  port(A, B: in std_logic; Sum, Cout: out std_logic);
end entity HALF_ADD;

architecture LOGIC of HALF_ADD is
begin
   Sum <= A xor B;
   Cout <= A and B;
end architecture LOGIC;
```

```
// Verilog
//
module HALF_ADD(A, B, Sum,Cout);
   input A, B;
   output Sum, Cout;

   assign Sum = A ^ B;
   assign Cout = A & B;

endmodule
```

(c) Source code

Figure 5-8. A half adder implementation in (a) VHDL, and (b) Verilog. The corresponding source code in VHDL and Verilog is shown in (c).

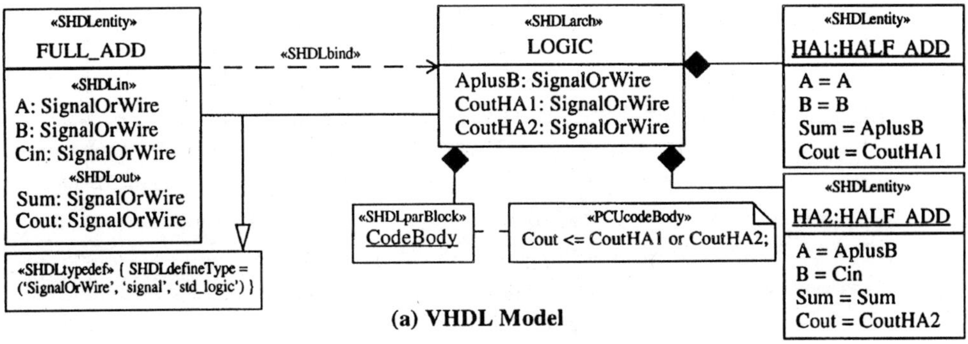

(a) VHDL Model

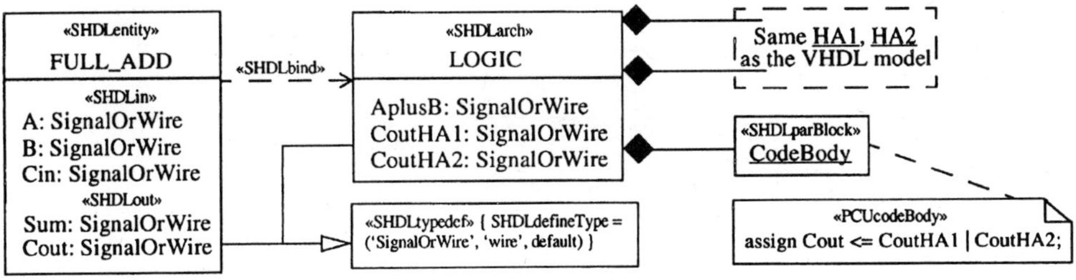

(b) Verilog Model

```
-- VHDL
--
entity FULL_ADD is
port (A, B, Cin: in std_logic;
      Sum, Cout: out std_logic);
end entity FULL_ADD;

architecture LOGIC of FULL_ADD is
  component HALF_ADD
    port (A, B: in std_logic;
          Sum, Cout: out std_logic);
    end component;
  signal AplusB, CoutHA1, CoutHA2: std_logic;
begin
  HA1: HALF_ADD port map (A=>A, B=>B,
        Sum=>AplusB, Cout=>CoutHA1);
  HA2: HALF_ADD port map (A=>AplusB,
        B=>Cin, Sum=>Sum, Cout=>CoutHA2);
  Cout <= CoutHA1 or CoutHA2;
end architecture LOGIC;
```

```
// Verilog
//
module FULL_ADD (A, B, Cin, Sum, Cout)
    input A, B, Cin;
    output Sum, Cout;

    wire AplusB, CoutHA1, CoutHA2;

  HALF_ADD HA1(.A(A), .B(B), .Sum(AplusB),
                     .Cout(CoutHA1));
  HALF_ADD HA2(.A(AplusB), .B(Cin),
                     .Sum(Sum), .Cout(CoutHA2));

  assign Cout = CoutHA1 | CoutHA2;

endmodule
```

(c) Source code

Figure 5-9. A full adder implementation in (a) VHDL, and (b) Verilog. The corresponding source code in VHDL and Verilog is shown in (c).

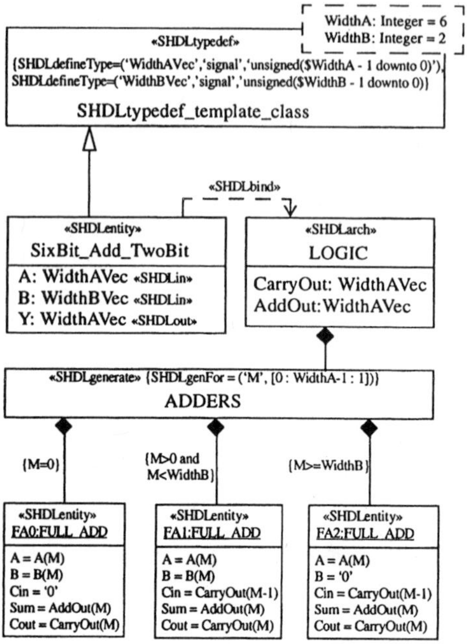


```
entity SIXBIT_ADD_TWOBIT is
  generic(WidthA: integer := 6;
          WidthB: integer := 2);
  port(A: in unsigned(WidthA-1 downto 0);
       B: in unsigned(WidthB-1 downto 0);
       Y: out unsigned(WidthA-1 downto 0));
end entity SIXBIT_ADD_TWOBIT;

architecture LOGIC of SIXBIT_ADD_TWOBIT is
  component FULL_ADD
    port(A, B, Cin: in std_logic;
         Sum, Cout: out std_logic);
  end component;
  signal CarryOut: unsigned(WidthA-1 downto 0);
  signal AddOut: unsigned(WidthA-1 downto 0);
begin
ADDERS: block
  begin
   for M in 0 to WidthA-1 generate
    if(M=0) generate
     FA0: FULL_ADD port map (A=>A(M),
       B=>B(M), Cin=>'0', Sum=>AddOut(M),
       Cout=>CarryOut(M));
    end generate;
    if(M>0 and M<WidthB) generate
     -- Port Map
    end generate;
    if(M>=WidthB) generate
     -- Port Map
    end generate;
end block ADDERS;
end architecture LOGIC;
```

(a) VHDL-targeted six-bit-add-two-bit adder **(b) Source code**

Figure 5-10. A six-bit-add-two-bit adder implementation in VHDL (a), and the corresponding source code (b).

5.6 Architecture Blueprint Modeling (ABprofile)

The UML profile for Codesign Modeling Framework (CMF) comprises of several subprofiles, as described in prior sections. The PCUprofile supplies for the other subprofiles the common facilities, as well as the essential link to the LPO. The EMprofile and IMprofile packages provide for a more convenient real-time application development by allowing exception handling routines, and interrupt service routines, respectively, to be modeled at a higher level of abstraction. On the hardware side of the Codesign Modeling Framework, the SHDLprofile addresses the need for UML modeling of synthesizable HDL languages, that, when applied in tandem with the other sub-profiles in the framework, renders a uniform hardware and software environment for the development of platform-centric SoC systems.

The Architecture Blueprint Modeling profile (ABprofile), on the other hand, deals with the UML representation of platform architectures that renders easy configuration and/or derivation of the desired target architecture.

5.6.1 Domain Viewpoint

A system platform can involve pre-designed and pre-characterized hardware, middleware, and software components (see Section 2.1, The Platform Concept, for further details). A combination of these components can result in myriad possible target architectures. As such, it is only appropriate that the platform architecture be represented abstractly in such a way that subsequent instantiations of platform-compatible components can conveniently occur that yield the target architecture as the desirable product. The abstract representation of platform architectures is specifically referred to in this book as an architecture blueprint, or simply blueprint (see Chapter 4: Definition 4.7, and Section 4.2.1.7).

In principle, the abstract architecture specified by an architecture blueprint aids the configuration/derivation of the target architecture by furnishing an architectural template for the system developer. A blueprint-specific tool or tools are expected to always accompany the blueprint to assist in the process of configuring/deriving the target architecture.

5.6.1.1 Modeling the Blueprint for Configuring/Deriving the Target Architecture

UML allows a hardware platform to be modeled as a Node. By associating components, applications and/or middleware with a Node, the developer can acquire the following:

Explicitly portray relationships among software and hardware components,

Draw pre-characterized Node information to be used for such tasks as performance analysis, and schedulability analysis.

However, in the hardware-software co-design environment, it often requires more than just representing hardware platforms and deploying software components in order to be useful. Routine co-design tasks such as mapping HDL code to a programmable logic device or instantiating a core onto the target architecture cannot be satisfactorily handled by means of standard UML notations. All these testaments mandate that a more comprehensive model shall be implemented to satisfy beyond the normal confines of software engineering. Now by associating a certain design with an architectural blueprint (AB) type, the developer and tool shall be able to interpret the appropriate relationship between them, some of which are shown in Table 5.4.

Table 5-4. Semantic inferrences of the relationships between design components and physical hardware

Design	Blueprint Type	Inferrence
HDL, EDIF netlist, Data	Programmable/Reprogrammable Unit, e.g. FPGA	Configuration file for a PRU unit is expected
HDL, EDIF netlist	Implementable Unit, e.g. ASIC.	Design must be fabricated
--None--	Drop-in Unit, e.g. peripheral devices, DSP	Hardware entities already in full-development; may require interface module for communication
General design component	Drop-in Unit: a processor	Design is deployed in a processor
Data	Memory Unit (MU)	Data resides in memory
Application; LPO component	Interface Unit (IFU)	Device driver, controller exists; Pin connections information exists.
General design component	Clock	Clock utilization
General design component	Timer	Timer utilization

The benefits of utilizing the AB types in the proposed approach can arguably be twofold: (1) as previously mentioned, implicative interpretation of the relationship between design and AB entities can be deduced, and (2) it furnishes coarse-grained categories in an architecture blueprint that can be used as search keywords within the LPO domain. Below is the list of these AB types:

- *Programmable/Reprogrammable Unit (PRU).* The category encompasses blueprint entities that can be programmed and/or reprogrammed on the field. Typically, an entity in this class provides a quick system prototype for the developer. Examples are Field Programmable Gate Array (FPGA), Programmable Logic Device (PLD), and Electronically Erasable Programmable ROM (EEPROM).

- *Implementable Unit (IU).* Affiliated blueprint entities in this class bear some similarities to those in the PRU category. They are designed by the developer, but must be sent for fabrication and packaging after the design and verification. Because of this distinct nature, the IU belongs in its own class apart from the PRU. Examples of AB entities in this category is the Application-Specific IC (ASIC), and standard cell devices.

- *Drop-In Unit (DIU).* The defining characteristic of AB entities in this class is that they have been pre-developed, i.e. all characteristics can be precisely acquired. Unless it is a processor, this type of entity has no direct interactions with the design and requires an interface unit (IFU), e.g. device driver and controller, to manage the communications. Example of AB entities in this category are peripheral devices, processors and coprocessors.

- *Interface Unit (IFU).* The entities in this class provides a communication means for the design and other AB entities. Examples of the IFU entities are PCI bus, Parallel I/O, and Firewire.

- *Memory Unit (MU).* The AB entities of this type represent the storage element class. Semantically, the MU is a subtype of the PRU; it exists primarily for the search purpose. Examples are RAM and ROM.

- *Clock.* The Clock type is a subtype of the DIU type. Nonetheless, because of its perennial presence in real-time systems, it merits to be distinguished from the rest of the class. When present, the «RTclock» stereotype in the UML real-time profile [29] shall refer to it.

- *Timer.* Like Clock, the Timer type is a subtype of the DIU and it corresponds to the «RTtimer» concept in the UML real-time profile [29].

5.6.2 UML Viewpoint

In defining UML extensions, i.e. stereotypes, tagged values and/or constraints, for the ABprofile package, the prefix AB is always attached to the names to differentiate them from similar or same names in other profiles.

5.6.2.1 Mapping Blueprint Domain Concepts into UML Equivalents

Physical hardware maps to a class that is stereotyped with «ABnode». A design unit, describing hardware implementation and software application alike, maps to a blueprint component represented by an «ABcomponent»-adorned class.

The «ABnode» stereotype comprises one tag: ABisKindOf. This tag permits the type of the «ABnode» class to be documented (see Section 5.6.1.1). The mapping of the «ABcomponent» class onto the «ABnode» class is specified by the «ABmap» stereotype adorned on a Dependency. Similarly, the «ABdeploy» stereotype is used to indicate the mapping of software application to the «ABnode».

5.6.2.2 Mapping the Blueprint Model Instance into the Physical Model

To prove the validity of using the AB model to configure/derive the target architecture in place of the physical model, the mapping that demonstrates the one-on-one relationship between the target architecture and the physical models is shown in Table 5.5.

From Table 5.5, it can be seen that the homomorphic mapping is possible with a couple of exceptions: there are no corresponding semantics in the physical model for the «ABprogram» and «ABbecome» stereotypes. This comes as an evidence for the inadequacy of the physical model as a modeling facility for the hardware-software co-design environment.

Table 5-5. Mapping of the AB's target architecture model into the physical model

Target Architecture Model	Physical Model	Target Architecture Model	Physical Model
'clock' Object	clock Node	'timer' Object	timer Node
Object	physical HW	parametrized Object	configured HW
Class Type	Node Type	Class Instance (owner)	Node Instance
Link	Communication	Class Instance (owned)	Component Instance
Dependency	Dependency	Composition	Composition
«ABdeploy»	«deploy»	«ABprogram»	--None--
		«ABbecome»	--None--

5.6.2.3 UML Extensions

To avoid any possible duplicate and ambiguity, all extensions defined in this profile are prefixed with AB.

«ABbecome»
This stereotype represents the relationship between a design unit and a hardware entity of type IU (see Section 5.6.1.1). It indicates that the design unit is a synthesizable HDL description of the hardware entity.

Stereotype	Base Class	Tags
«ABbecome»	Dependency	--None--

The following constraint is defined for this stereotype:

- The client element at the tail of the arrow must be of IU type.

«ABcomponent»
This stereotype models a design unit that may indiscriminately represents application software, or a hardware component (see Section 5.6.1.1).

Stereotype	Base Class	Tags
«ABcomponent»	Class	--None--

	Object	

«ABdeploy»
This stereotype infers that a design unit can be supported by a hardware entity. Normally, it signifies the residency of a design unit in the hardware. It is the ABprofile equivalent of the standard «deploy» stereotype.

Stereotype	Base Class	Tags
«ABdeploy»	Dependency	--None--

«ABnode»
This stereotype represents hardware as view by the architecture blueprint (see Section 5.6.1.1).

Stereotype	Base Class	Tags
«ABnode»	Class Object	ABisKindOf

The tag for this stereotype is defined as follows:

Tag Name	Tag Type	Multiplicity	Domain Concept
ABisKindOf	Enumeration: ('pru', 'iu', 'diu', 'ifu', 'mu', 'clock', 'timer')	0..1	AB Type (see Section 5.6.1.1)

«ABprogram»
This stereotype represents the relationship between a design unit and a hardware entity of type PRU, where the design unit is used to program the hardware entity (see Section 5.6.1.1).

Stereotype	Base Class	Tags
«ABprogram»	Dependency	--None--

The following constraint is defined for this stereotype:

- The client element at the tail of the arrow must be of PRU type.

5.7 UML to SystemC Mapping

In Section 2.4, the collaborative usage model of the platform-centric and the SystemC approaches was presented, and potential advantages were discussed. This section focuses on the mapping between the two models to demonstrate that models coherence can be maintained, and that the mapping process can be automated.

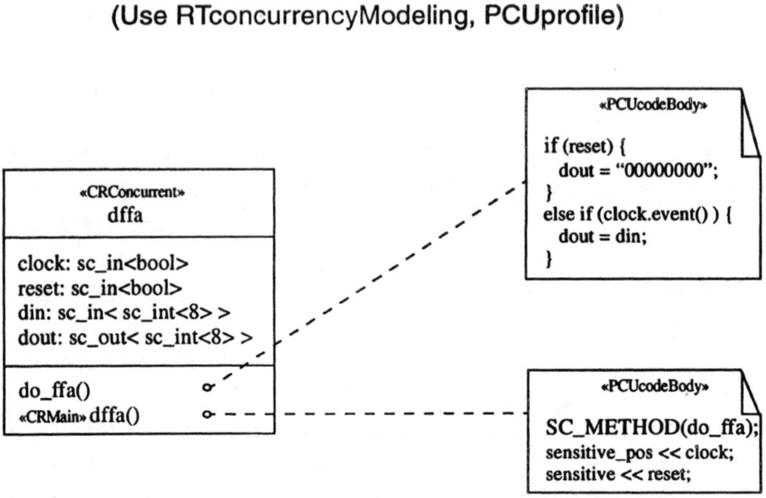

Figure 5-11. A SW-viewed UML/SystemC model of an 8-bit D-F/F

Because SystemC is entirely based on an OO language, i.e. C++, mapping to and from UML is quite natural and intuitive. However, of particular interest is how specialized SystemC macros and functions map to the UML constructs. Table 5.6 summarizes this mapping from the software and hardware points of view. The fact that the proposed approach is software-biased probably would make the software modeling scheme the more attractive between the two. This SW-viewed UML/SystemC model utilizes the RTconcurrencyModeling package defined in the UML Real-Time Profile [29] as the basis for capturing SystemC specialized constructs. On the other hand, the hardware approach could be useful for migrating existing UML/HDL models to the UML/SystemC. In such a case, an intelligent tool would be capable of understanding the SystemC constructs from the SHDLprofile-based UML models, resulting in minimal changes required of the UML models. It also eases the task of replacing HDL code with the corresponding SystemC.

Table 5-6. Mapping of the SystemC constructs to the platform-centric UML models (SW and HW perspectives)

SystemC	UML (SW Perspective)	UML (HW Perspective)
SC_MODULE	«CRConcurrent»	«SHDLentity»
SC_CTOR	«CRMain»	«SHDLarch»
SC_METHOD	method	method
SC_THREAD	«CRAction» {isAtomic=true}	«SHDLprocess»
SC_CTHREAD	«CRAction» {isAtomic=true}	«SHDLprocess» {SHDLsensitive=('clock')}

sc_main()	«CRAction» sc_main	«SHDLentity» sc_main

(Use SHDLprofile, PCUprofile)

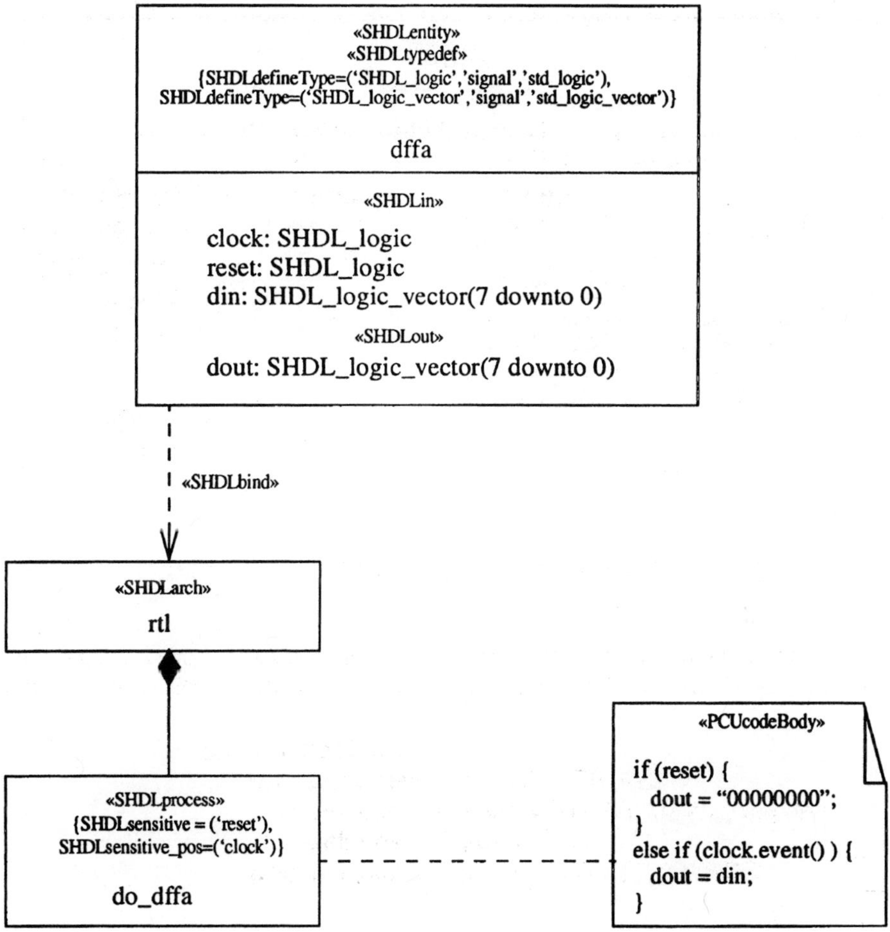

Figure 5-12. A HW-viewed UML/SystemC model of an 8-bit D-F/F

As an example, consider an implementation of a D flip flop with an asynchronous reset, whose VHDL description is presented below. Figure 5.11 delineates the SW-viewed UML/SystemC model of such a flip flop, while Figure 5.12 depicts the model as perceived from the HW viewpoint.

```
-- VHDL implementation of an 8-bit D-F/F
-- dffa.vhd

entity dffa is
      port( clock  : in   std_logic;
            reset  : in   std_logic;
            din    : in   std_logic_vector(7 downto 0);
            dout   : out  std_logic_vector(7 downto 0) );
end dffa;

architecture rtl of dffa is
begin
      process(reset, clock)
      begin
            if reset = '1' then
              dout <= "00000000";
            elsif clock'event and clock = '1' then
              dout <= din;
            end if;
      end process;
end rtl;
```

Chapter 6

DESIGN CASE STUDY: A DIGITAL CAMERA

6. INTRODUCTION

This chapter demonstrates the applicability and benefits of the proposed platform-centric SoC design approach by using it to develop a simplified digital camera system. A digital camera is a complex system comprising both mechanical and electronic components—rendering it very well-suited as an application case study for this book. The chapter starts by giving an overview of typical digital camera operations, as well as relevant mechanical and electronic components. Thereafter a set of general requirements is given, and the CMF development process begins in the manner prescribed by this book in earlier chapters. After detailing the tasks involved in each main step of the platform-centric development process flow, i.e., the platform-independent, platform-analysis, platform-dependent, and system derivation process steps, the chapter concludes by comparing cost-effectiveness of the proposed approach utilizing parametric cost estimation models, such as COCOMO II.2000.

6.1 Digital Camera System

Comparing to a traditional film camera, a digital camera operates very much on the same principle, although minute operational details differ considerably. It has by and large the same user interface as that of a film camera, but with additional options only attainable through digital photography. A digital camera captures and stores images in digital format on a storage device. Figure 6.1 depicts a block diagram of a typical digital camera system.

6.1.1 Image Acquisition Module

Just like a traditional film camera, light reflecting off the scene or subject is directed by the lens onto the image acquisition module, which normally comprises an image sensor, and/or an analog to digital converter (ADC). The duration and amount of light exposure is regulated by a shutter and an aperture, respectively.

An image sensor constitutes the core for any digital camera system. It is based predominantly on either the Charge Coupled Device (CCD) or Complementary Metal Oxide Semiconductor (CMOS) technology (see Table 6.1 for comparison). Within these sensors, photocells are arranged in rows and columns, and become electronically charged when exposed to light. "This charge can then be converted to an 8-bit value where 0 represents no exposure while 255 represents very intense exposure of the cells to light [120]." Some of the columns of the photocell array are covered with a black strip of paint, and are used for zero-bias adjustments, i.e. white balancing, of all the cells in the array. Because the CMOS imager is inherently noisy, an ADC is often integrated to help prevent further image quality deterioration [121].

Normal digital camera operations commence with the process of determining proper settings for the scene or subject to be photographed. Such tasks typically involve adjusting the focus, setting image quality, measuring and gathering shooting parameters, and selecting an appropriate shutter duration and aperture opening (f-stop in photographic term). Once the required parameters are set and the shutter button is pressed, the following sequence of operations typically ensues [121]:

- The shutter is closed; the sensor becomes temporarily inactive, and is instantly flushed off all residual charges. This step is to prepare the sensor to capture a new image.

- Depending on the camera and the settings, the residual charges that are flushed off the sensor may be analyzed to acquire the proper settings for automatic point-and-shoot operations. Or if a LCD viewfinder is present, it will display the flushed image on the screen.

- The sensor becomes active and, at the same time, the shutter opens, exposing the sensor to light—charging it as a result. The shutter remains open for the specified exposure duration, before closing again. The image can now be captured and streamed off to the Image Conditioning module.

- The shutter re-opens, and the camera is ready to take another picture.

Table 6-1. Comparison of the CMOS and CCD image sensors [121, 122]

Criteria	CMOS image sensor	CCD image sensor
Cost-effectiveness	More cost-effective	Less cost-effective
Image Quality	Less superior (more susceptible to noise)	More superior (less susceptible to noise)
Pixels Access/ Download	Direct access possible (similar to memory access)	No selective read. Must serially flush off whole image
Integrated ADC	Yes	No

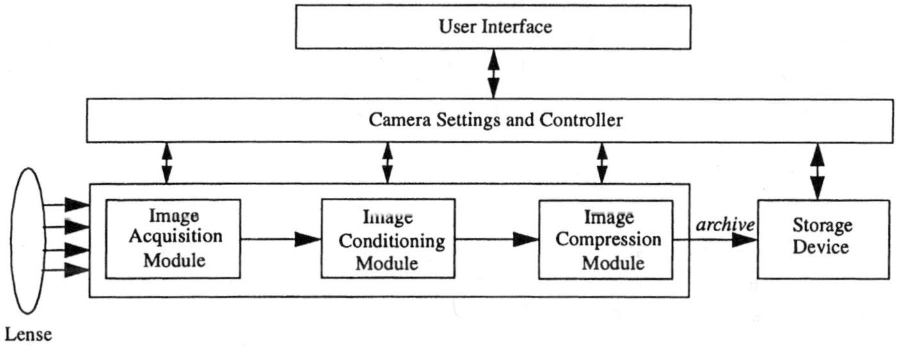

Figure 6-1. Block diagram of a typical digital camera system

In many digital cameras, the time it takes for the sensor to flush, as well as to read and set the shooting control parameters is often non-trivial—ranging from as little as 60 milliseconds to as long as 1.5 seconds [121]. This so-called shutter lag can be improved by utilizing a larger buffer memory, and/or a faster processor. Higher communication bandwidth can also speed up the lag. In some CMOS image sensors where rows and columns can be selectively read, the shutter lag tends to improve as well.

6.1.2 Image Conditioning Module

The functionality of this module chiefly concerns color processing and image enhancing tasks aimed to render a visually better image for the user.

The actual algorithms for achieving such a result vary from one camera to another, depending on the imposed design criteria.

When the Image Conditioning module receives the digital data representation of the image from the Image Acquisition module, specifically, the ADC, it first determines whether or not de-mosaicing is required. Most digital cameras today carry only one image sensor, instead of multiple image sensors for multiple color components. As such, to be able to acquire a full resolution image from a single image sensor, the color filter array (CFA) architecture is utilized to assign a color component to each photocell [121]. The CFA-imposed digital data from this sensor can later be de-mosaiced with respect to the CFA architecture to derive the full resolution image. An example of the CFA architecture is the popular Bayer pattern.

Once a full resolution image is acquired, color processing and image enhancing tasks can begin. These tasks involve, for example, color balance and saturation settings, white balancing, noise removing, as well as other image effects and enhancement such as sharpness enhancing, red-eye eliminating, or sepia coloring.

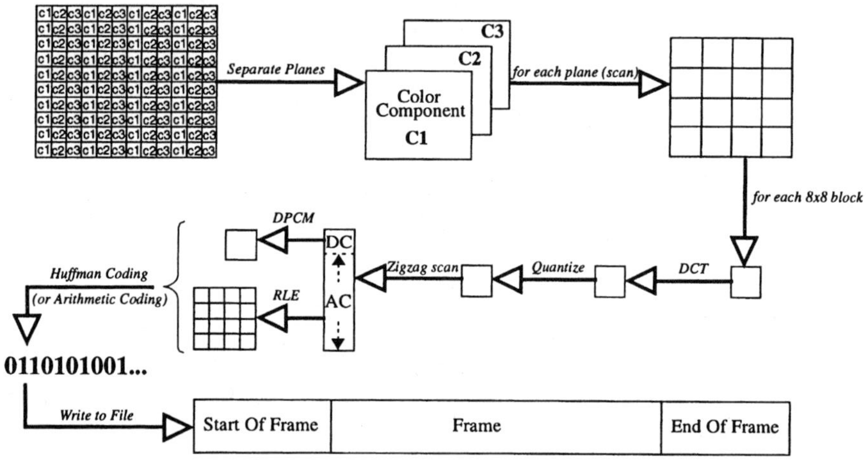

Figure 6-2. Block diagram of the baseline JPEG encoder

6.1.3 Image Compression Module

To make optimal use of the storage device as well as to expedite upload/download time, the preprocessed image from the Image Conditioning module may further be compressed before writing onto the media. Most digital cameras let the user choose the image quality settings that determine (1) the compression quality, and (2) the image dimension. Often, this module is also responsible for generating image thumbnails.

Arguably, the most popular image compression algorithm used by digital cameras today is the JPEG standard. JPEG works well with continuous-toned images, e.g. natural scenes or photograph pictures. It defines both lossy and lossless algorithms, as well as the embodiment of minimum requirements, called Baseline JPEG or simply just JPEG, that guarantees portability across different decoder implementations. The baseline JPEG is lossy; it, nonetheless, permits different compression quality settings to be specified—for example, a value in the range of 0 to 100 where 0 means most compressed, worst quality, and 100 least compressed, best quality. Figure 6.2 illustrates the baseline JPEG encoder block diagram.

In Figure 6.2, the JPEG encoder takes the color components representing the image as its inputs, and divides each of them into non-overlapping blocks of 8x8 pixels. For each block, JPEG transforms the spatial data into frequency domain using the Discrete Cosine Transform (DCT). Then in the quantization step, it rounds these frequency data (DC and AC components) to the closest pre-defined values in the quantization tables, attempting to minimize the number of total bits—hereby, resulting in a smaller image size. As the human vision is less sensitive to high-frequency components, larger gains in compression ratio may further be achieved by allowing larger errors to occur and/or eliminating some high-frequency components altogether. The quantized data are then entropy-coded using Differential Pulse Code Modulation (DPCM) and Huffman encoding for DC components, and Run-Length encoding (RLE) and Huffman encoding for AC components. Finally a JPEG file can be produced using the predefined file format.

6.2 Digital Camera System Requirements

For the case study, a functional prototype of a digital camera is developed that must meet the following initial requirements. As before, requirement levels are indicated using the keywords described in the guideline furnished by the Internet Engineering Task Force (IETF)'s RTF2119 [63], as shown in Table 4.1.

6.2.1 Functional Requirements

Functional requirements describe system behaviors. For the case study, the following functional requirements are imposed.

6.2.1.1 General Operational Requirements

The image acquisition and conditioning modules, illustrated in Figure 6.1, are assumed to be implemented by another team. This simplified digital camera shall comprise only the user interface, the JPEG encoder module (see Figure 6.2), and the archiving step that writes the compressed image onto a media. It must provide means to upload images onto a PC. It is also responsible for implementing the camera control logics.

6.2.1.2 User Interface's Operational Requirements

This simplified digital camera system has a pre-set shutter speed, as well as fixed aperture and focus settings. However, it permits the user to select image quality (normal or good), as well as a single shot or 2-shot burst mode. The camera status should be appropriately displayed.

Stored images can be removed via the digital camera's user interface. An image upload operation causes all stored images to be transferred to a PC, without deleting the images on the media; it is analogous to the file copy operation.

6.2.1.3 Input and Output

The inputs are the color components of the image; how these inputs are read shall be determined by the actual implementation of the JPEG encoder. The eventual output is a compressed JPEG file or files (in a 2-shot burst mode) written onto a media.

6.2.2 Non-functional Requirements

Non-functional requirements are requirements that do not concern system behaviors; they encompass such metrics as:

- Performance, e.g. time required to process an image,
- Size, e.g. number of logic elements, and/or size of embedded software,
- Power, e.g. measure of average power consumption,
- Energy, e.g. battery lifetime.

Vahid and Givargis [120] describes non-functional requirements as consisting of *constrained* metrics, and optimization metrics. Constrained metrics are values that must not violate a specified threshold(s); whereas, optimization metrics refer to certain design goals that aim to improve the system. This section describes the non-functional requirements for the case study.

6.2.2.1 Operating Time Constraint

To be useful, the time used to take and store one image shall not exceed 1 second. The time is measured from the moment the shutter is pressed to the moment the camera is ready to take another picture.

Although faster operating time is normally desirable, it must also be thoroughly justified.

6.2.2.2 Heat Dissipation and Energy Requirement

Even though the prototype is implemented on the NiOS development board that relies on an AC-to-DC adaptor as the power source, the actual product will likely use batteries. Therefore, it is desirable to minimize the power consumption as much as possible, so as to prolong battery life. It is required, however, that no cooling fan shall be used.

6.2.2.3 Hardware Platform Requirements

The functional prototype of the simplified digital camera system being developed for the case study shall be deployed on the Altera NiOS soft core embedded processor platform targeted for the EP20K200EFC484-2X programmable logic device (PLD). Compatible platform components may be utilized that are available in the LPO, such that the resultant product closely emulates the actual digital camera operating environment. Descriptive summary of the NiOS processor and platform, as well as the EP20K200EFC484-2X PLD device is presented as follows (see their respective datasheets from Altera [81] for more details).

NiOS Embedded Processor

The NiOS embedded processor is a general-purpose, five-stage pipelined RISC soft processor core in which instructions run in a single clock cycle. The NiOS CPU can be configured for a wide range of applications; its 16-bit instruction set is targeted for embedded applications in particular. The NiOS supports a fully-synchronous address/data bus interface. In addition, it features a configurable 16-bit or 32-bit data path, 64 vectored interrupts, 1-to-

31-bit single-clock shift operations, as well as 128 to 512 general-purpose registers. Two hardware-assisted multiplications are available that permit a single bit per cycle multiplication (MSTEP) or a fast integer multiplication (MUL). Instruction set extensions can be realized via the custom instruction feature. Figure 6.3 depicts the 32-bit NiOS processor.

NiOS Platform

The NiOS embedded processor provides for customizable, on-chip peripherals that attributes to a convenient configuration of the NiOS platform. Once properly configured, the developer can have the peripheral bus module (PBM) for such peripherals as timer, SRAM, FLASH, universal asynchronous receiver/transmitter (UART), and parallel I/O (PIO) automatically generated. The following PBM features are fully customizable:

- Base address and address span
- Data width
- Read/Write access restrictions (read-only, write-only, read-write)
- Wait states, and
- IRQ signal/priority.

Figure 6.4 depicts the NiOS platform that portrays the communication between the NiOS embedded processor and its peripherals.

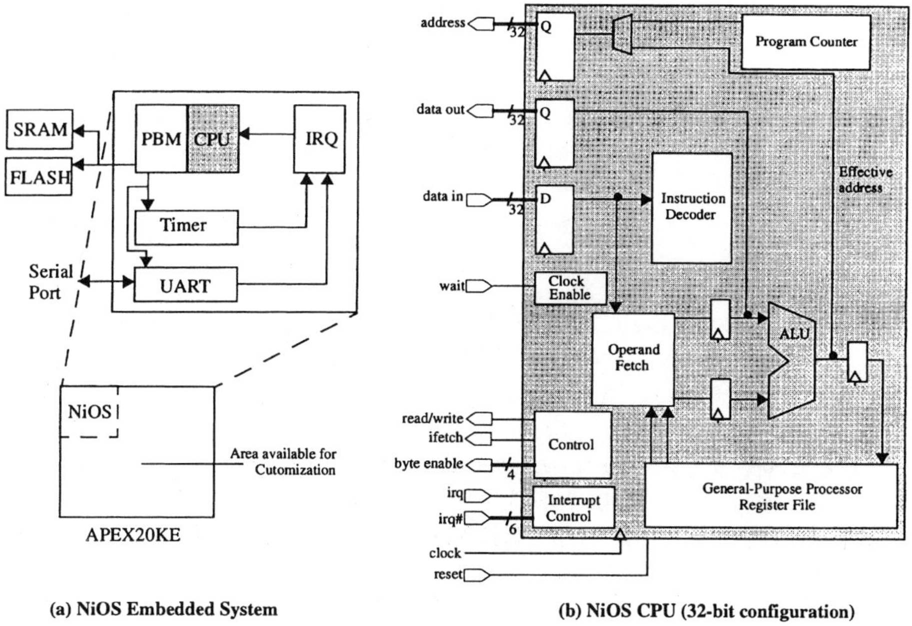

Figure 6-3. The NiOS embedded processor [81]

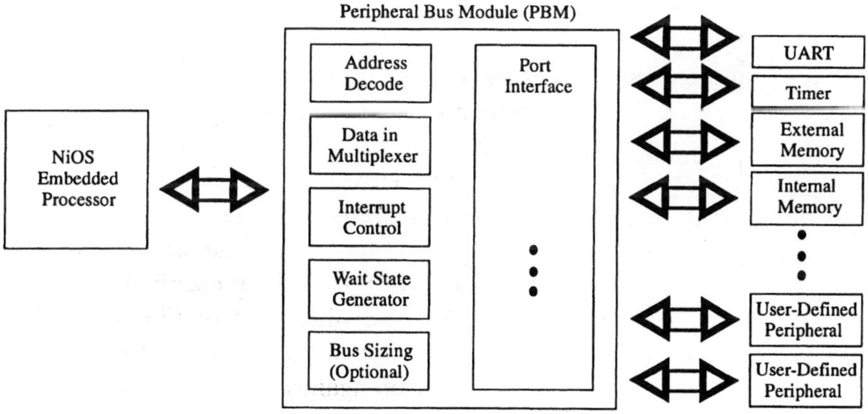

Figure 6-4. NiOS platform, showing communication between the NiOS processor core and its peripherals [81]

EP20K200EFC484-2X PLD Device

The EP20K200EFC484-2X belongs in the Altera APEX20KE device group within the APEX20K family. Like all of its sibling devices, the

APEX20KE is built upon the MultiCore architecture consisting of logic array blocks (LABs), whose basic units are the logic elements (LEs). Each LAB comprises 10 LEs; 16 LABs can be combined to form a new hierarchical structure called MegaLAB. In addition to these 16 LABs, a MegaLAB contains an advanced embedded structure called an embedded system block (ESB).

The ESB is the heart of the MultiCore architecture. Each ESB contains 2048 programmable bits that can be configured as:

- product-term logic, which is superior for control logic functions such as address decoding and state machines,
- LUT-based logic, or
- three types of memory: dual-port RAM, read-only memory (ROM), or content-addressable memory (CAM) which allows the address to be identified from a data input.

Specifically, the EP20K200EFC484-2X device contains 52 ESBs/8320 LEs, with the maximum RAM bits (ESB bits) of 106496. It is housed in the 484-pin fine line, ball grid array (BGA) package; the maximum of 376 I/O pins are available for use by the system developer. The device allows up to two phase lock loop (PLL) implementations, and supports four voltage interfaces: 5 V, 3.3 V, 2.5 V, and 1.8 V.

The NiOS embedded processor core is optimized for the APEX20K programmable logic devices. When programmed on the EP20K200EFC484-2X, the 32-bit configuration uses approximately 20% of the resources (approximately 13% in the 16-bit configuration), and can execute up to 50 MIPS with the fastest permissible clock speed of 50 MHz.

By using the platform-centric approach to develop the simplified digital camera described above, the required tasks follow the prescribed stages that are treated in detail in Chapter 2 and also illustrated in Figure 6.5. The remainder of this chapter presents specific work details for each stage, before concluding with the cost comparison against the SpecC methodology [20].

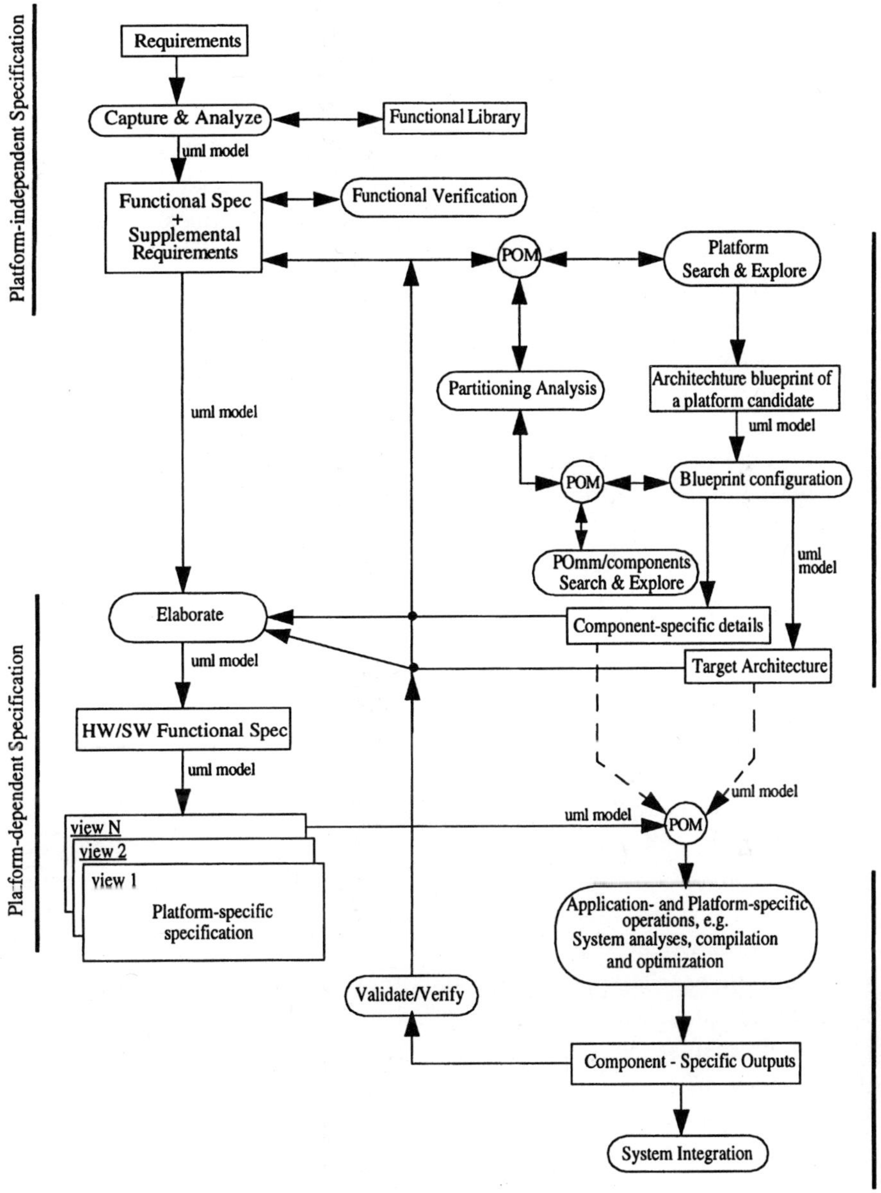

Figure 6-5. The platform-centric SoC method design flow, CMF, as proposed in this book.

6.3 Platform-Independent Specification

This stage of the proposed platform-centric approach involves deriving the functional specification of the system that is still independent of any platform specifics. Due to this nature, proven techniques in software engineering are applicable to be used by the system developer. Some well-known UML processes include the Rose™ Unified Process [102], the UML/Catalysis as described by Graham [123], and the Rapid Object-Oriented Process for Embedded Systems (ROPES) covered in detail in the UML book by Douglass [28].

Much like the object-oriented programming techniques that specify functional interfaces without committing prematurely to any specific implementation, the platform-centric SoC design method imposes no specific UML process on the derivation of this specification. As a result, the proposed approach is flexible, and can readily adapt to the diverse requirements of the complex development process of the real-time embedded SoC systems. To derive the platform-independent specification in this stage, the book employs the modified version of the Requirement Specification Process as detailed in Maciaszek [36]. This Use Case driven technique commences with the task of requirement determination that clarifies the initial requirement document, shedding more light to it from the developer's point of view. It then enters the elaboration loop from either the Use Case analysis or Class analysis entry points (or both, in parallel), and iteratively refining, and deriving the desired specification using various UML techniques. The resultant UML model comprises the static system structure captured by the Class diagram, the system behavior represented by such a diagram as Sequence or Collaboration, and the system state as described by the State diagram. Figure 6.6 depicts this process flow.

The requirements determination task often requires multi-lateral efforts involving many subtasks that will elicit, negotiate, validate, manage, and model the customer's requirements (see [36], [123] for details). The requirements document resulted from these subtasks is subsequently analyzed that eventually will yield the platform-independent specification.

It is quite normal in practice that the initial requirements are ambiguous and/or incomplete. The requirements statement presented in Section 6.2 decidedly provides insufficient information for the system developer to proceed further into the development process, and the developer needs to go through the requirements determination process (preferably together with the customers). Since the requirements determination task is somewhat beyond the scope of this book, its detailed procedures will be omitted. The resultant extended requirements document, which is the product of reworking the requirements statement in Section 6.2, appears as follows:

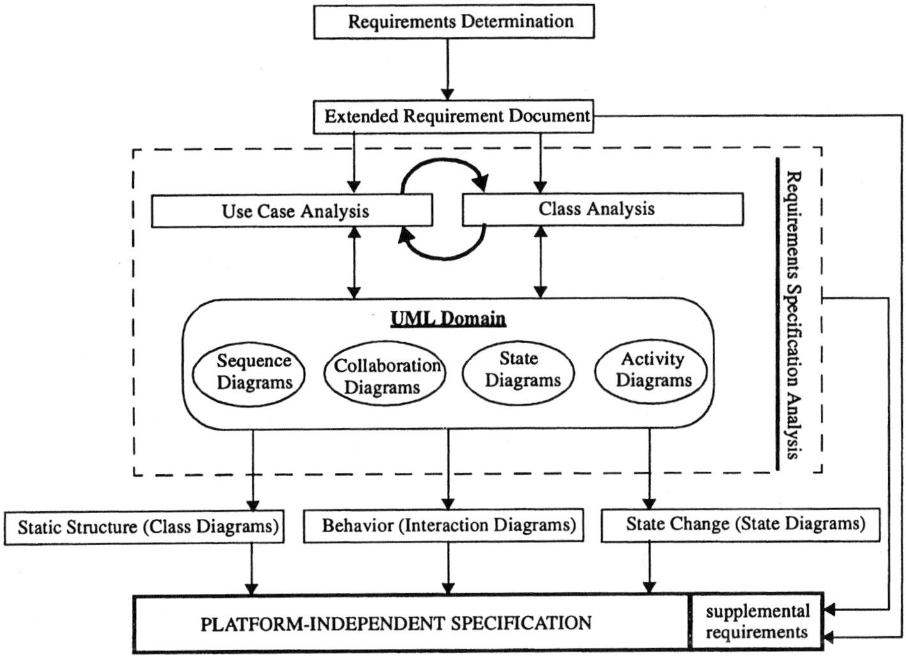

Figure 6-6. Platform-Independent Specification Process Flow

Digital Camera's Extended Requirements

(R1) The user powers on the camera. The camera displays a ready message and also an image count. A LED light illuminates. The camera is ready to take a picture.

(R2) The user presses the shutter. The camera takes a picture and stores the image in the memory. During the operation, the LED light goes off and comes back on when finished. The image count on display gets incremented. The camera displays a ready message.

(R3) The user enters the menu mode by pressing the menu button. The user can browse the available options that include: image quality setting, shot mode, upload, and delete.

(R4) The user uses the select button to show and to change the current setting for each available option. For the image quality setting, the user can choose between "good" or "normal", with "normal" as the default. For the shot mode, the user can select either single-shot mode (default) or two-shot burst mode. The upload and delete settings can be either "yes" or "no" (default).

(R5) The user uses the done button to commit to the settings and brings the menu view up one level iteratively until exiting the menu mode altogether, where the camera is ready to take a picture.

(R6) The user uses the done button to activate the upload operation. When upload operation is requested, the camera transmits all images stored in the memory to the PC, where the interface software writes the JPEG files from these data. No image is deleted upon completion. When finished, the status is displayed with no change in the image count and the camera is back in the menu mode.

(R7) The user uses the done button to activate the delete operation where the camera removes all images from the memory. The image count on display is reset to zero, and the camera is back in the menu mode.

6.3.1 Use Case Analysis

Depending on the entry point that the extended requirements document enters the Requirements Specification Analysis domain, the Use Case analysis may either be the kick-start or the follow-up activity of the string of requirements specification refinement tasks that proceed sequentially and iteratively. It is also possible, and arguably beneficial, that the two analysis processes be performed simultaneously so as to impart one's analytical strength onto the other, and ergo, becoming more effective in deriving the platform-independent specification.

The Use Case diagram models the system as seen looking in from the outside, as such, making it attractive for modeling the requirements document which generally describes the user's viewpoint towards system functionalities. In UML terms, a Use Case diagram comprises one or more use cases, each of which captures a system behavior that is outwardly visible to an actor or actors, and that readily responds to external events caused by the actor. A Use Case diagram is a graphical representation of use cases and actors and how they interact.

Use cases can be derived from the identification of tasks of the actor [36]. From the digital camera's extended requirements, two actors are identified, namely, a user, and a PC, and seven use cases are manifest in the requirements statement. Figure 6.7 portrays the Use Case diagram of the digital camera system that is derived directly from the extended requirements. An extension point in a use case merely allows a functionality of another use case to be summoned at a specified location.

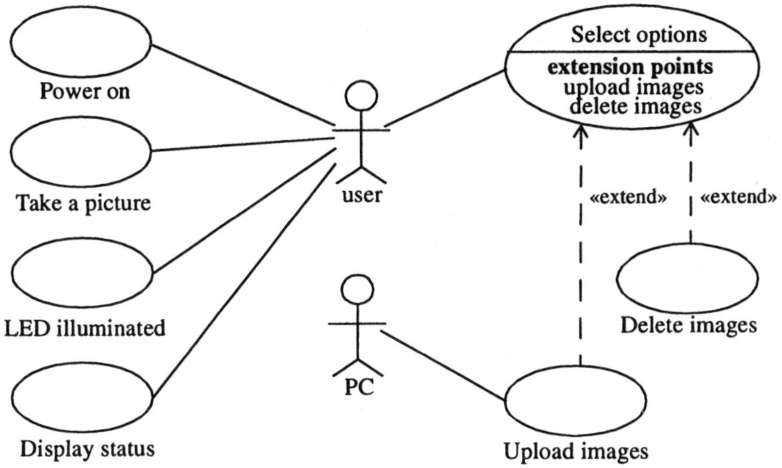

Figure 6-7. Initial Use Case diagram as derived directly from the digital camera's extended requirements.

From this *initial* Use Case diagram, each individual use case, and its respective actor, gets further elaborated to produce a more detailed requirements document which narrows the focus down from the system to the use case level. Details and formats of this derivative requirements document vary from one organization to another, but normally include information such as pre-/post-conditions, and the main and alternative flow, as demonstrated below for the *Take a picture* use case.

Use Case Document 1: Take a picture

Use Case:	Take a picture
Description:	This use case allows the user to take a picture using the digital camera
Actor:	User
Pre-conditions:	(PRE-1) The camera powers on without error. (PRE-2) It is in a ready state.
Main flow:	(MR-1) The use case begins when the shutter is pressed. The LED light goes off. (MR-2) The camera system reads in the input color components. The shot-mode setting determines the number of times these color components need to be read per one press-shutter operation. (MR-3) The camera system compresses the input color components using the baseline JPEG algorithm. The compression quality parameter is determined by the "GOOD" or "NORMAL" setting from the setting menu. (MR-4) The camera system stores the compressed image (or images) in the memory. It also increments and displays an image count. The LED light comes back on.
Alternative Flow:	No Alternative flow.
Post-conditions:	(POST-1) The compressed image is stored in the memory so that it can later be retrieved/deleted. (POST-2) The number of stored images is updated. (POST-3) The system state is unchanged. (POST-4) The shutter comes back to its non-pressed state and ready to be pressed again.

This *Take a picture* requirements document can further be analyzed and captured using the Activity diagram so as to delineate the logical action states of the main and the alternative flow—the task which could prove helpful in eliciting more requirements. The resultant Activity diagram could also be used as a reference for the Class Analysis task. From the Take a picture requirements document, the following action states are educed: Read color components, Compress input data, Store compressed image, and Update image count. Figure 6.8 portrays the Take a picture Activity diagram.

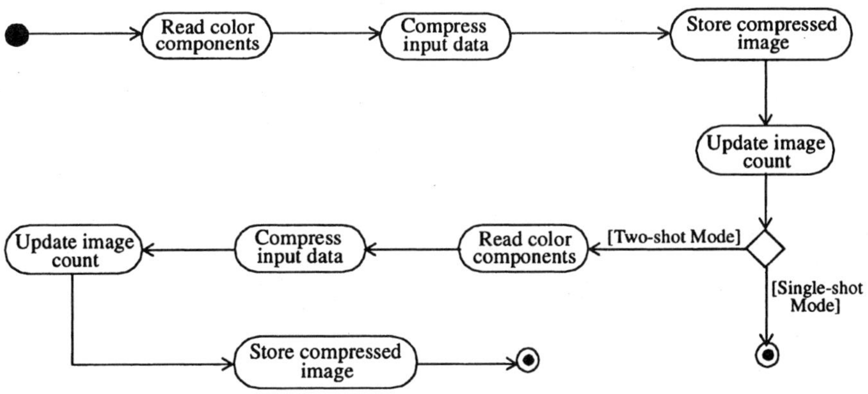

Figure 6-8. The "**Take a picture**" Activity diagram

By iteratively performing such analyses for all the initial use cases, additional use cases manifest that can contribute to a more profound insight about the system. In so doing, the developer may proceed:

- depth-first, until no additional requirements can be derived for a particular use case, at which point the developer moves on to a new use case, or
- breadth-first, where the use cases are discovered layer by layer, or
- alternatively, between Use Case and Class analyses, treating the two analysis tasks as concurrent processes that are supportive of one another.

When developing an embedded system within the platform-centric (or generally, co-design) environment, the Use Case diagram that captures the extended requirements typically will reveal most of the peripheral components to the developer. The subsequent Use Case analysis on a peripheral-related use case tends to uncover a communication interface between software and hardware components in the system. Consider the Use Case analysis of the Take a picture requirements document. It is apparent from the flow statements that the only actor for this use case model is the shutter. Then, by looking at the system from the actor's (shutter) point of view, it is easy to conceive that none of the activities being described in the requirements document can possibly be done directly by the actor. A new use case is required that handles the stimulus generated by the actor (see

MR-1), and that behaves as a proxy to the other activities in the requirements document, as depicted below. The *Use Case Document 2* that follows details the requirements for the Handle signals use case.

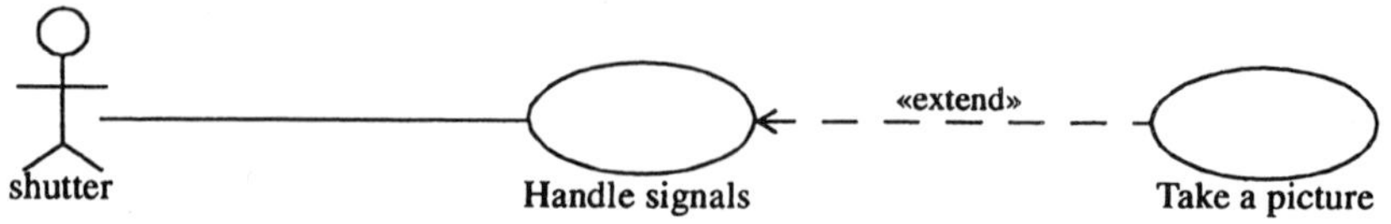

Use Case Document 2: Handle signals

Use Case:	Handle signals
Description:	This use case intercepts the shutter signal and activates the system to take a picture
Actor:	Shutter
Pre-conditions:	(PRE-1) The camera system is ready to accept a signal from the actor. (PRE-2) The actor is ready to send a signal (PRE-3) All pre-conditions in the *Take a picture* use case apply, i.e. the power is on safely and the system is in the ready state.
Main flow:	(MR-1) The use case begins when the shutter is pressed. (MR-2) A signal is sent to the camera system. (MR-3) The system intercepts the signal, examines it. (MR-4) [At the extension point] the system begins taking a picture once it is certain that the signal comes from the shutter.
Alternative Flow:	(AR-1) If the signal does not come from the shutter, appropriate extended use case is called.
Post-conditions:	(POST-1) The compressed image is stored in the memory so that it can later be retrieved/deleted. (POST-2) The number of stored images is updated. (POST-3) The system state is unchanged. (POST-4) The shutter comes back to its non-pressed state and ready to be pressed again.

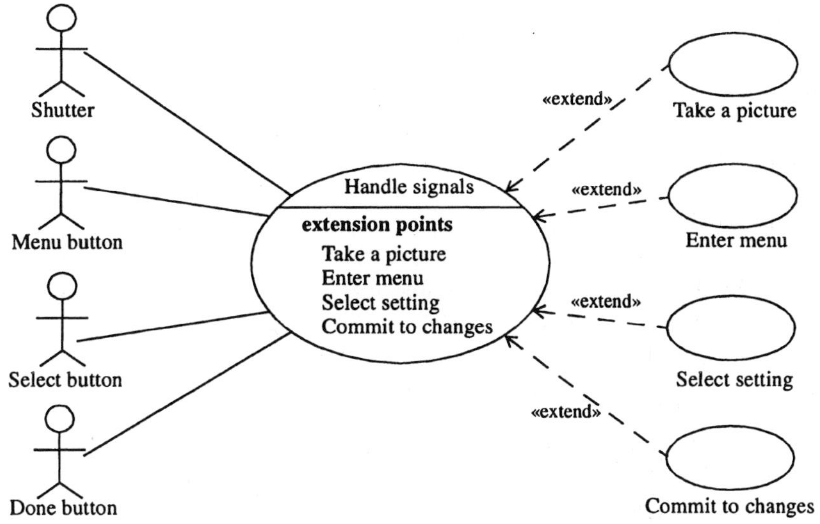

(a) The *Handle signals* Use Case diagram

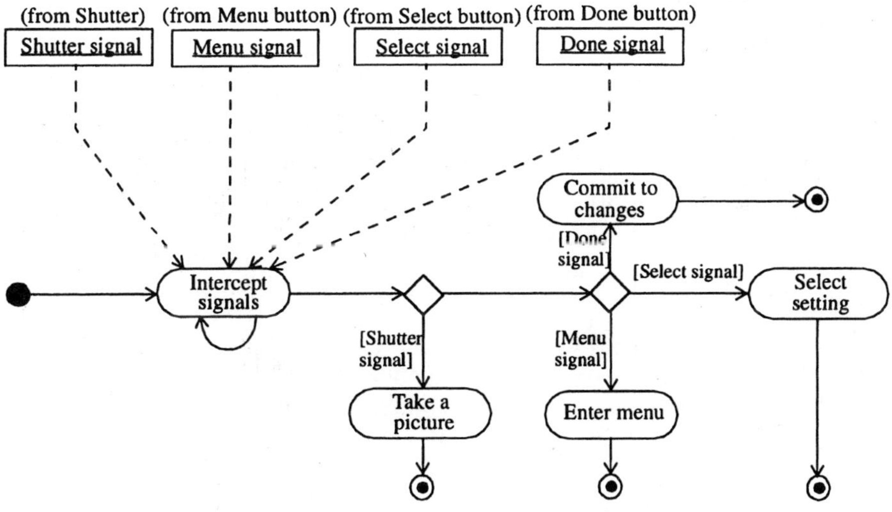

(b) The *Handle signals* Activity diagram

Figure 6-9. Derived from the **Handle signals** requirements document, (a) the eventual Use Case diagram, and (b) the Activity diagram

It is also evident from the Alternative Flow statement AR-1 that there exist other use cases that interact with the Handle signals use case as well.

As the matter of fact, upon completion of the Use Case analysis, the Use Case diagram, and the respective Activity diagram will appear as depicted in Figure 6.9.

6.3.2 Class Analysis

Although this book presents the Class analysis task after the Use Case analysis, such an order is not necessarily true in practice. Depending on the practicality of how the requirements document enters the Requirements Specification domain, the developer may perform the Class analysis first, or choose to run both analysis tasks simultaneously.

A handful of techniques exist in books, papers and the Internet, that can contribute to the Class analysis process. Douglass [98], in particular, treats this ever-mystical topic comprehensively and in great detail. A few other useful books dealing with this subject matter include, but not limited to, those by Graham [123], Douglass [28], Maciaszek [36], and Fowler [25].

Where the Use Case analysis derives the platform-independent specification as it is outwardly visible to the actors, the Class analysis task does just the opposite: It attempts to derive the specification as it is seen internally. The Class analysis task presented herein employs the Noun analysis technique to extract candidate classes from the requirements document. UML diagrams, especially the Sequence and the Use Case diagrams, and Class analysis techniques such as the Abbott's textual analysis [124] and the Class-Responsibility-Collaboration (CRC) Cards technique, which has been developed by Beck and Cunningham (see URL http://c2.com/doc/oopsla89/paper.html for the original paper, or the book by Wirfs-Brock, Wilkerson, and Wiener [125] for further details) can be utilized to help identify class attributes and methods, as well as relationships among the candidate classes.

The guideline to the Abbott's textual analysis, as presented in [123] is summarized in Table 6.2. The "noun analysis" simply is a subset of this technique that concerns only with improper nouns that infer candidate classes. By grouping candidate classes with respect to some criteria, the developer can break down the problem into smaller chunks in a divide-and-conquer manner, which can be handled more easily and effectively when also applying other techniques to help identify class attributes and methods, as well as potential relationships among candidate classes.

6.3.2.1 Noun Analysis/Textual Analysis

The noun analysis refers to the process of selecting improper nouns from the requirements document that can potentially be characterized as a class in the UML model. Performing such a task as part of the iterative analysis process to derive the platform-independent specification, this book analyzes the nouns in the extended requirements document, before focusing on each individual use case, and proceeding in a depth-first manner that eventually produces a complete use-case-centric Class diagram as the result. The system's Class diagram is derived by merging all of the use-case-centric Class diagrams together, performing additional analyses only where necessary during the merge. In the extended requirements and Take a picture requirements, the candidate nouns are underscored.

In order to traverse depth-first, consider again the Take a picture requirements. By grouping the candidate nouns from this document and the extended requirements, basing the grouping criteria on the Take a picture use case, the following candidate classes are identified:

- camera system
- LED light
- color components
- display
- [baseline] JPEG
- shutter
- compressed image
- memory
- [take a picture] operation
- user

Table 6-2. Guidelines for the Abbott's textual analysis.

Part of Speech	UML Model Component	Example
proper noun	instance	the EP20K200EFC484-2X PLD device
improper noun	class/type/role	push-button switch
doing verb	operation	start
being verb	classification	is a
having verb	composition	has a
stative verb	invariance condition	own
modal verb	data semantics, pre-condition, post-condition, or invariance condition	must be
adjective	attribute value or class	good, normal
adjectival phrase	association, operation	back in the menu mode
transitive verb	operation	enter
intransitive verb	exception or event	goes off, commit to

By concentrating on these nouns (classes), and doing the Abbott's textual analysis on them within the confines of the Take a picture use case, additional class characteristics are acquired as shown below:

Text	Functions	Potentially owned by	Potentially associated with
- *image count*	attribute	camera system; memory	--N/A--
- *image quality setting*	attribute	display; camera system; operation; JPEG	--N/A--
- *shot mode*	attribute	display; camera system; operation; JPEG	--N/A--
- *ready message*	attribute	display	--N/A--

- *store*	operation	memory	compressed image
- *press*	operation	shutter	user
- *increment*	operation	memory; camera system	operation
- read (input)	operation	color components	operation; JPEG
- *read shot mode setting*	operation	display; camera system; operation; JPEG	JPEG
- read image quality setting	operation	display; camera system; operation; JPEG	JPEG
- *display*	operation	display	camera system; memory; operation
- *on/off*	operation	LED light	camera system; operation
- take a picture	operation	operation	--All Classes--

Although not evident from the textual analysis, it is fairly intuitive to perceive the camera system class as maintaining an ownership relationship with all other classes in the system—hence, inferring composition in the Class model. Figure 6.10 depicts the early development of the Class diagram as derived from the Noun analysis/Abbott's textual analysis above.

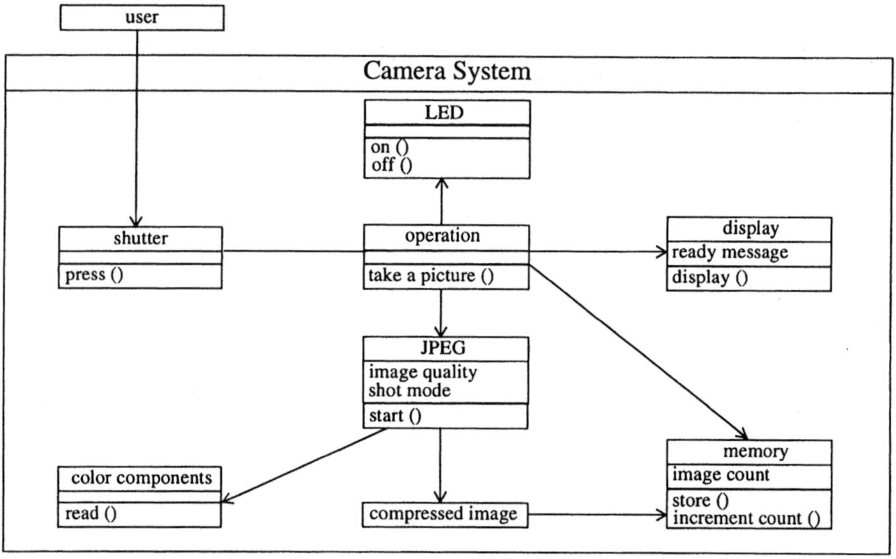

Figure 6-10. Preliminary use-case-centric Class diagram derived from the Noun Analysis/Textual Analysis

6.3.2.2 *Code Reuse*

Advanced knowledge of the NiOS platform utilization requirement gives the developer a certain degree of prescience that could impact the decision making process. Surfacing almost naturally is the choice of using the C programming language and/or NiOS macro code to develop the application software, as well as the opportunity to reuse the public-domain JPEG library (in C) to expedite the development process.

The developer who is familiar with the JPEG standard would likely know also that its success has come substantially from the courtesy of the Independent JPEG Group (IJG) for their effort in distributing free and highly portable C-code implementation of the algorithm. The platform-centric approach envisages the C JPEG library as legacy software that could be encapsulated using the UML notations as shown in Figure 6.11. The embellished «PCUrun» stereotype specifies the Makefile to be executed—permitting the inclusion of the library at the modifiable source code level rather than at the stationary archive level (e.g. jpeglib.a).

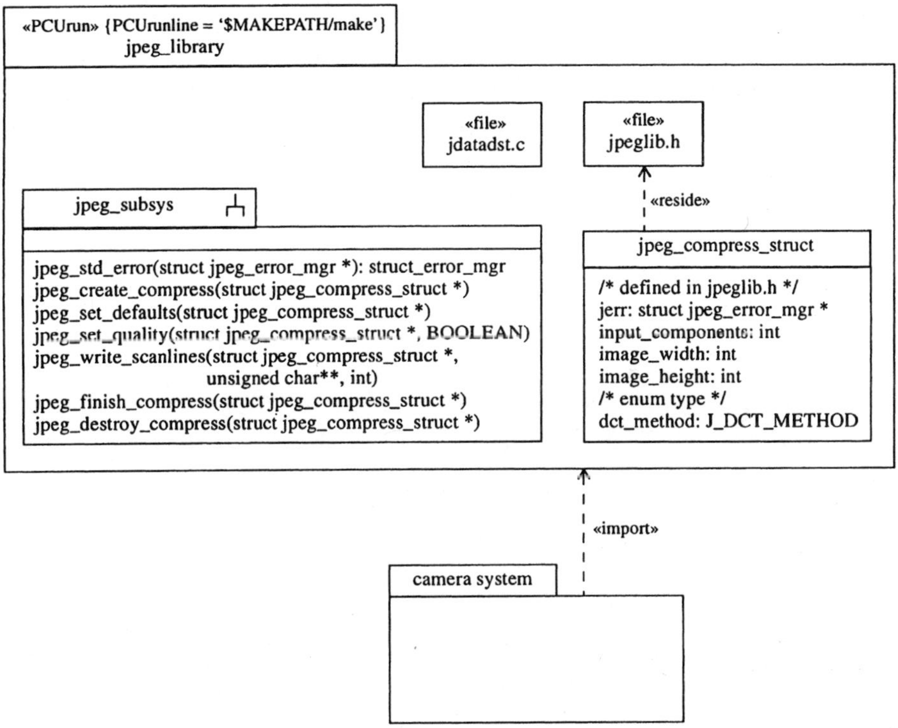

Figure 6-11. UML-encapsulated JPEG library. The figure shows the subsystem package that provides a functional interface to the required library functions

The IJG's JPEG library imposes further requirements that affect the format and choice of the input color components, as well as the structure of the compression engine that includes the image quality attribute. The compressed image class identified earlier can be implemented by the C structure jpeg_compress_struct. The memory class, renamed media to reflect the non-volatile nature of the memory, will be used from either the library or the camera system packages. The image quality attribute corresponds to the compression quality concept and can be set by calling the jpeg_set_quality() function. Figure 6.12 depicts the Class diagram, which presently includes the services from the JPEG library package. In the figure, the JPEG class is also renamed encoder to reflect the fact that it only copes with the forward operation of the JPEG standard.

By identifying roles and responsibilities of the classes in Figure 6.12, and then modeling their interactions using the interaction diagrams, that comprise the Sequence and the Collaboration model, the developer can potentially elicit additional information which could manifest itself as class attributes or operations, or spawn new requirements that merit their own Use Case and/or Class analysis.

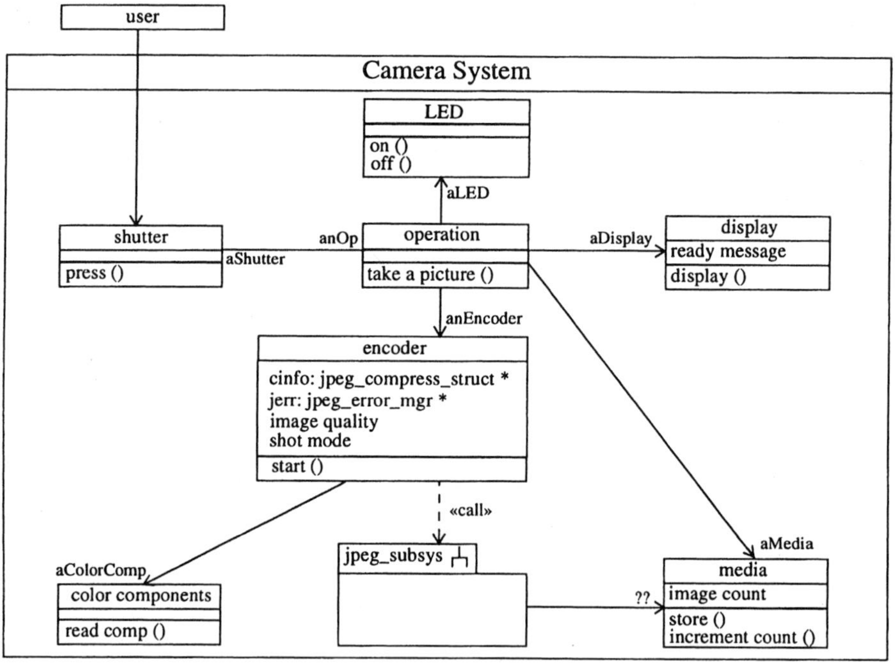

Figure 6-12. UseCase-centric Class diagram utilizing the IJG's JPEG library package

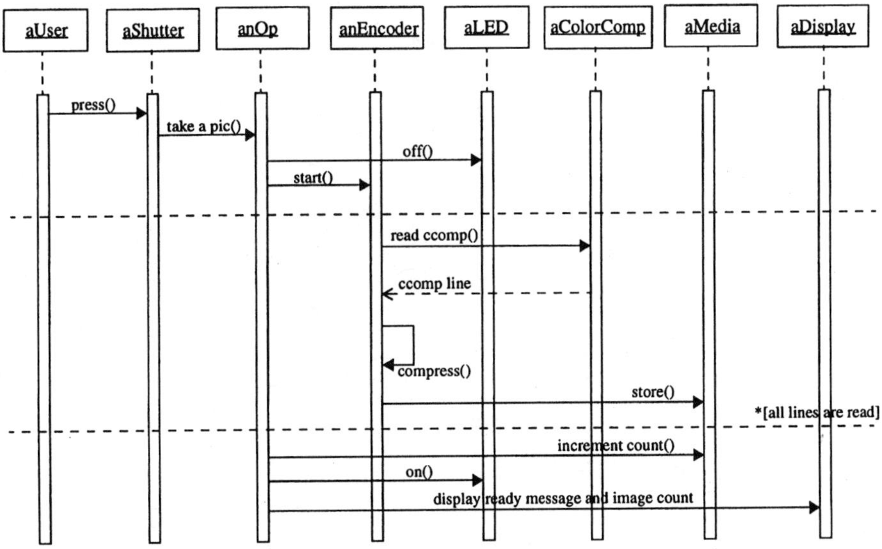

Figure 6-13. Sequence diagram describing the main scenario for Figure 6.12

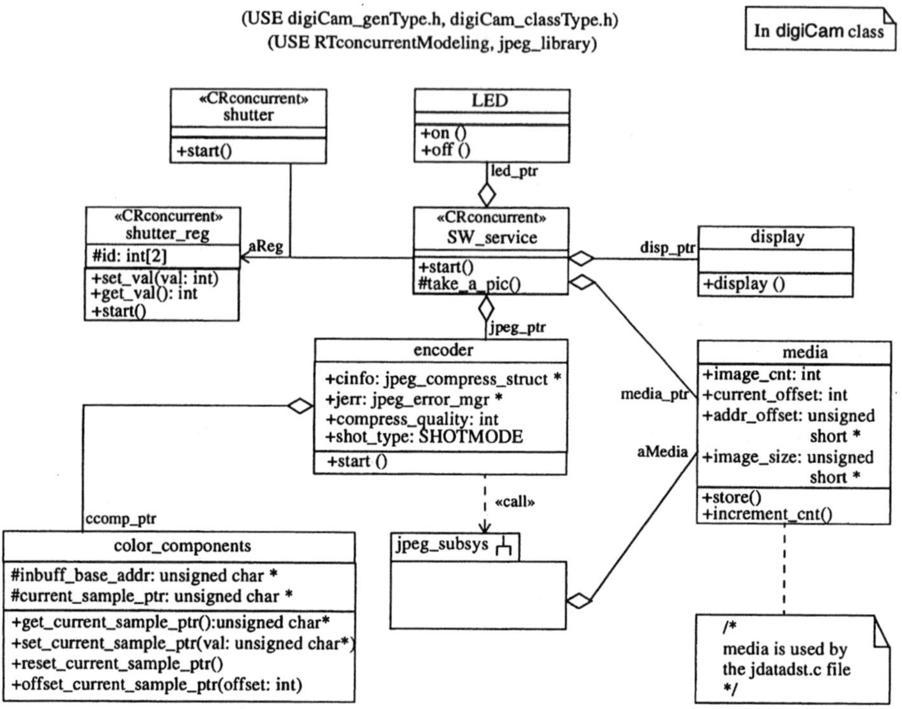

Figure 6-14. Detailed Class diagram for the **Take a picture** use case

The Sequence diagram in Figure 6.13 portrays a scenario that describes the principal collaboration of the classes in Figure 6.12. As evident from the figure, an awkward relationship between the shutter and the operation class necessitates a closer scrutiny. The thorough analysis of their relationship would likely result in an identification of an interface class that is perennially active, waiting for the shutter signal, and that is also responsible for furnishing a proper response to the intercepted signal—the idea resembling the interrupt handling mechanism concept as depicted in Figure 6.9.

The *read comp (),* which returns an input scan line (*ccomp line*), the *compress ()*, and the *store ()* functions all come from the library, and must be renamed accordingly. The IJG's JPEG library also already specifies for the developer the scan-line format. Information such as the need to keep track of the stored image locations and the current memory location can become more obvious via the diagrammatic representation. Likewise, the behaviors that may raise doubt, e.g. "Ought the increment *count ()* operation to be called by the operation class or by the media class?", or "Does the media class actually play two distinct roles?", tend to become more distinguishable, graphically, as well. Figure 6.14 illustrates the detailed resultant Class diagram of the Take a picture use case.

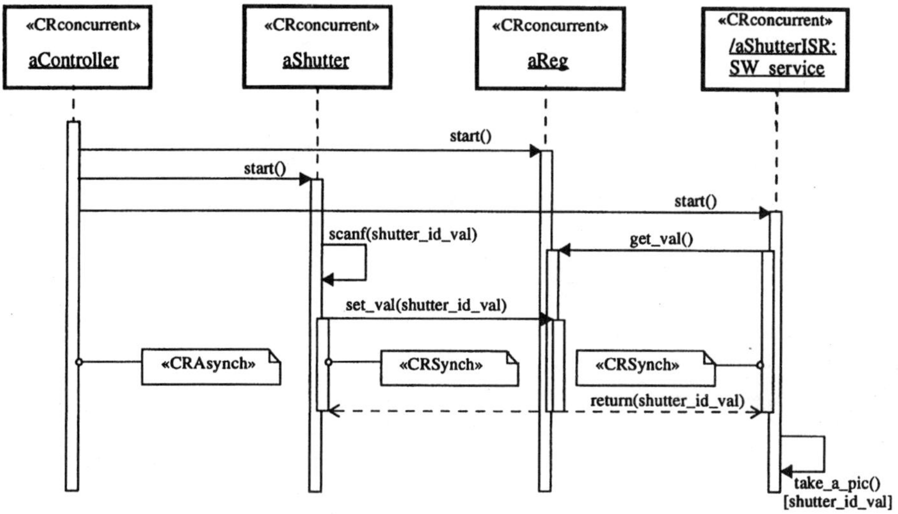

Figure 6-15. Sequence diagram describing the action that leads to an activation of the take_a_pic() function

A few notable changes are evident in Figure 6.14. The operation class is more properly renamed SW_service (SW for switch) and the camera system class now becomes the digiCam class that possesses the main function. The digiCam employs the concurrency modeling facility (the RTconcurrentModeling package [29]) to capture the concurrent characteristics of the shutter, shutter_reg, and SW_service classes. The press() function is replaced with the start() function that internally simulates the press action as coming from the standard input, i.e. a keyboard. The Sequence diagram in Figure 6.15 delineates the detailed behaviors of these concurrent classes, as seen by a controller class that is not shown here in the figure. The C-like pseudo code of the start() functions appears below.

Pseudocode of the start() function

The shutter_reg class

```
void start():
  pipe(id);
```

The shutter class

```
void start():
  if(fork == 0):
```

```
for(;;):
  scanf(shutter_signal_id);
  aReg.set_val(shutter_signal_id);
```

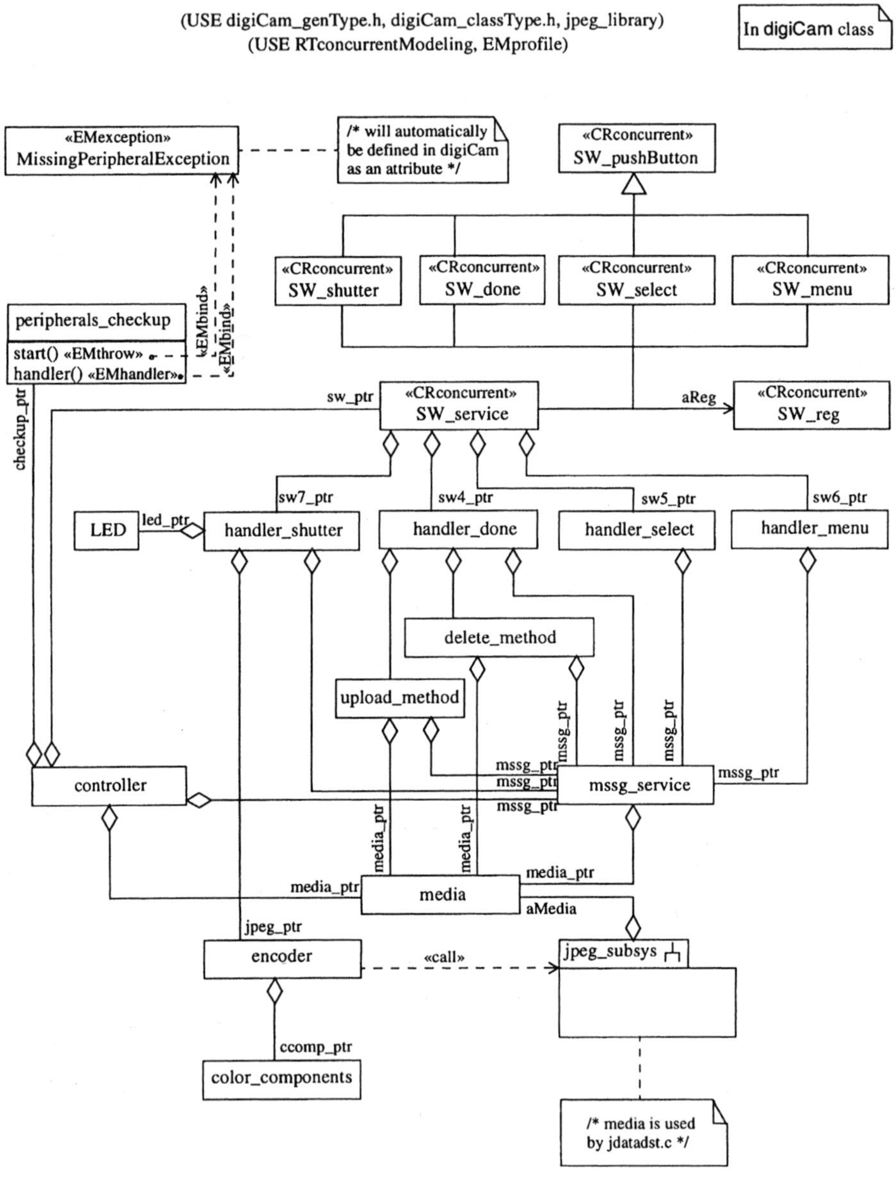

Figure 6-16. Detail-minimal Class diagram for the digital camera system

The SW_service class

```
void start():
  for(;;):
    aReg.get_val();
    if (shutter_signal_id)
      take_a_pic();
```

6.3.3 Concluding Remarks

To derive the platform-independent specification for the simplified digital camera system, this book performs the Use Case and Class analyses in a cyclical manner, resulting in an iterative refinement process that yields the desirable specification as the outcome.

As previously demonstrated, the adopted requirements analysis approach commences by identifying the primary use cases and classes from the initial requirements document. Then, for each primary use case, it determines the relevant classes, including the primary classes already identified, before proceeding with the Use Case and Class analyses, in a depth-first manner, until the final UseCase-centric Class diagram results. The platform-independent specification is derived by merging all such Class diagrams together, while performing further analyses as necessary.

Figure 6.16 depicts the space-optimal Class diagram that, along with the Sequence and the State diagrams, as well as the supplemental requirements document, constitutes the platform-independent specification.

Although not shown explicitly, all the classes in Figure 6.16 belong to the digiCam class (via the composition). The SW_shutter, SW_menu, SW_select, and SW done are the generalized classes of the SW_pushButton, all of which share the same device register (SW_reg) that allows the SW_service class to intercept the incoming press signal, and call the appropriate service routine. The display class is also renamed mssg_service to be more specific about its characteristics. Another notable addition is the utilization of the EMprofile package (see Chapter 5, Section 5.3 for details), which is discovered during the Power-up Use Case analysis. For simplicity, this camera only checks the system readiness by querying the existence of all anticipated peripherals—the operation that is performed by the peripherals_checkup class during power-up.

Mindful readers might notice that the platform-independent specification presented herein perceives the system development process as encompassing both hardware and software, permitting the hardware components and architecture to be configured as well as designed. This is to contrast with the current UML modeling practice where hardware is often regarded as being external to the development process and the hardware components and architecture can only be configured to model the relationships between hardware the software system under development. The specification is also reflective of the anticipated hardware/software system structure and behaviors, the benefits of which could be twofold:

- It provides for the developer the precise functional specification of the hardware components expected to be acquired, configured or designed for the system, and

- It eases the task of transitioning from the platform-independent specification to the HW/SW specific platform-dependent specification, as shall be seen later in this chapter.

Thus far this section has covered a lot of ground demonstrating, for the first stage of the proposed design flow as depicted in Figure 6.5, how the platform-centric SoC design method utilizes the cross-disciplinary techniques in Software and Requirements engineering to derive the platform-independent specification from the initial requirements document. To conclude, this book presents the supplemental requirements document as follows.

Digital Camera's Supplemental Requirements

(NF-R1) Let timeA be the time when the shutter is pressed, and timeB be the time when the camera is ready to take another picture. Then it is required that when taking a picture of NORMAL quality:

$$timeB - timeA \leq 1\ second,$$

for at least 95 percent of the time.

Note: The number of input samples, and effectively the image sensor capacity, is still unknown and must be determined.

(NF-R2) Faster is better.

(NF-R3) The final product will operate on battery, as such it should operate with as little power consumption as practically possible.

(NF-R4) No fan allowed.

(NF-R5) Must use the NiOS soft core, and the EP20K200EFC484-2X APEX20KE device

(NF-R6) To be competitive, throughput of 1 Mega-samples/second, corresponding to a 640x480-pixel, color image sensor, is expected.

(NF-R7) [From the Take a picture Use Case] To be operational, the power must be on and remains on, and all required peripherals must exist and functional, i.e.

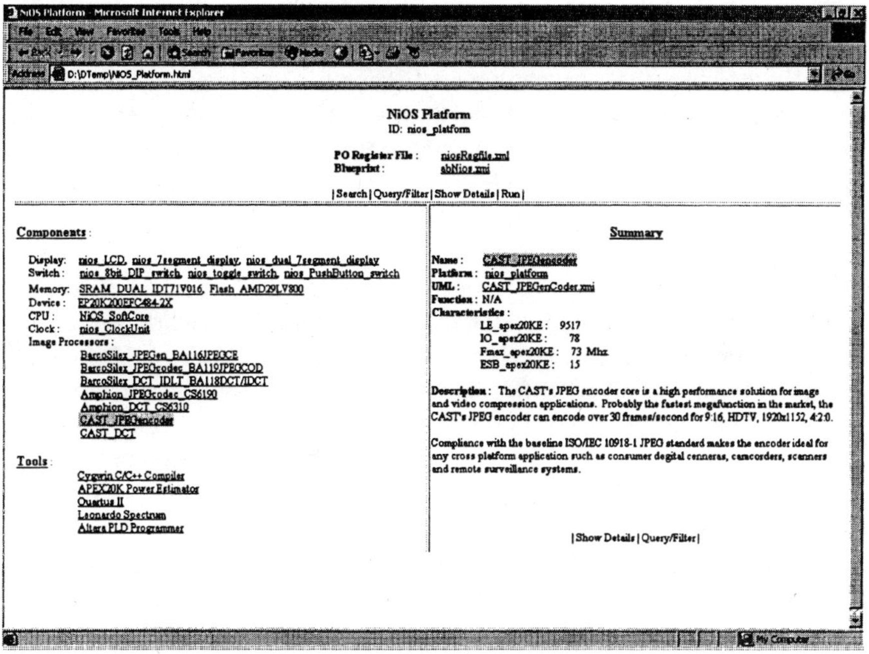

Figure 6-17. A simple POM interface window portraying the NiOS platform, with a short summary on the JPEG Encoder module provided by CAST Inc. (http://www.cast-inc.com)

```
context digiCam:
    inv: (self.current_state = STATE_READY or
          self.current_state = STATE_MENU) implies
          self.power_state = ON
    pre:  self.current_state <> STATE_ERROR
```

6.4 Platform Analysis

The principal task in this stage involves identifying the platform from within the LPO that is best suited the requirements represented by the platform-independent specification. Such a task is non-trivial. In Figure 6.5, this task is illustrated as an iterative process involving the search and exploration of available platforms in the LPO, as well as the hardware/software partitioning analysis of the platform-independent specification per each candidate platform. In most hardware/software SoC design approaches, where there exists no explicit support for the derivation process of the platform-independent specification, the equivalent of this stage is where the process flow commences, whose input is the requirements specification[1] which has been derived elsewhere.

The search and explore task requires the developer to interact extensively with the LPO via the Platform-Object Manager (POM) interface software. A simple POM, like the one shown in Figure 6.17, relies on XSL/XSLT to format the LPO's XML database and presents it to the developer in a familiar HTML format. The Summary section provides the developer with quick cues as per the characteristics of each platform members (POmm/components and POmm/tools). While the Search operation permits non-LPO searches, an operation such as the XPath-based Query/Filter helps the developer zero in on a certain aspect of the LPO database, making it useful for the task at this stage. To simulate the work environment anticipated by the POM, the DTD schema documents and the XML register files have been placed in different machines, accessible only through the Internet. The information applicable to all POmm/components and POmm/tools is amassed from the actual NiOs components and tools, furnished by their respective providers. Due to time restriction, only a simple POM is developed for the case study. It is envisaged, however, that a more sophisticated version of the POM software, possibly the one that integrates itself seamlessly into the platform-centric UML CASE tool, would contribute favorably to the efficiency of the proposed approach.

For this particular case study, actual work is curtailed owing largely to the specific requirements for the NiOS platform and the EP20K200EFC484-2X PLD device. In a more typical scenario, the developer would have to explore the available choices of platform, and assess and compare the feasibility of using each one of them for the problem at hand. This process can be automated using CAD tools. It can also be achieved manually, or by utilizing both approaches together. The use of platforms helps to allay the difficulty of such a task by placing well-defined constraints, i.e. the platforms and their respective components, onto the design space—contracting the sheer size of it as a result. Anyhow the developer's experiences still contribute considerably to the success of this task. Figure 6.18 depicts the architecture blueprint of the Altera's NiOS platform system, where the constraints and enumeration types are presented in Figure 6.19.

[1] Requirements specification is a generic term referring to the same concept as the platform-independent specification, which is a platform-centric jargon.

6.4.1 Automated Architecture Selection and System Partition

Algorithms exist that can assist the system developer in partitioning the system as well as in selecting the target architecture. Axelsson provides a comprehensive overview on a number of such algorithms in his book [23], the detailed treatment on some of which can be found in [91, 92, 93, 94].

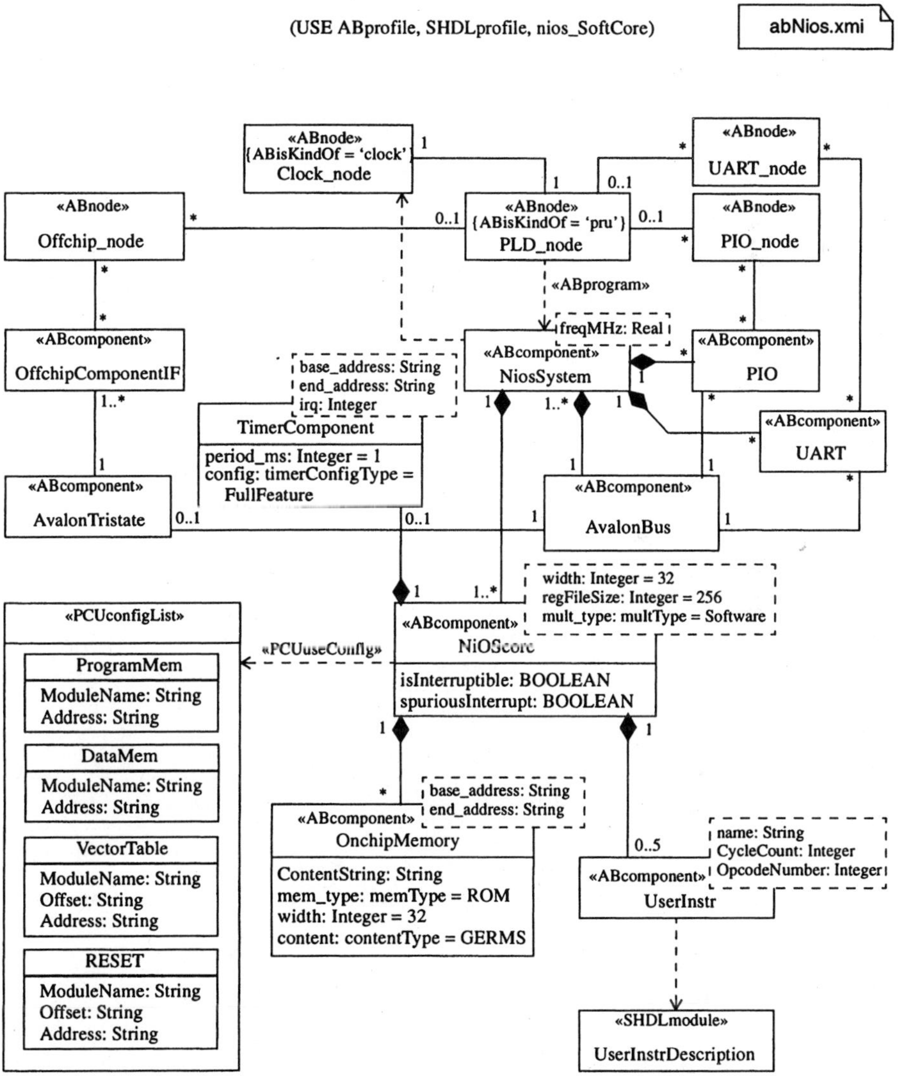

Figure 6-18. The Architecture Blueprint of the NiOS platform (abNios.xmi), depicting the platform structure

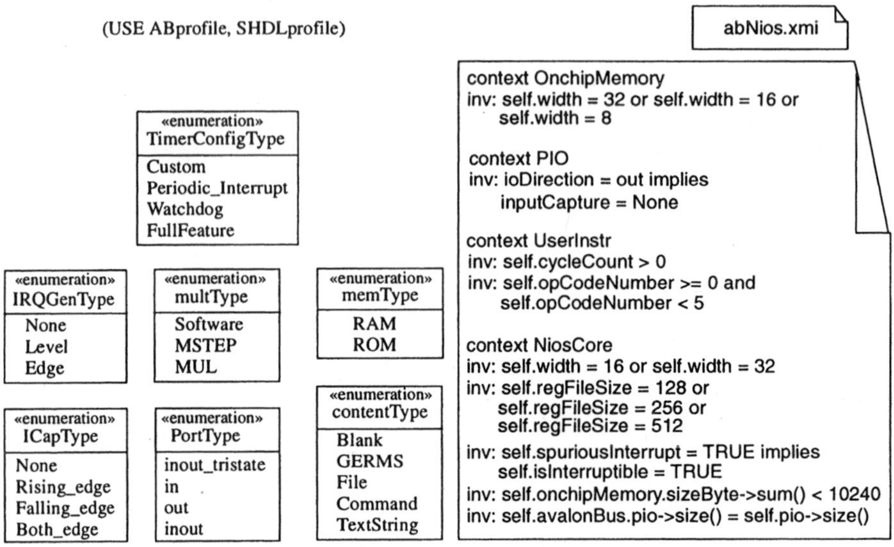

Figure 6-19. The Architecture Blueprint of the NiOS platform (abNios.xmi), depicting a partial list of constraints and enumerated types

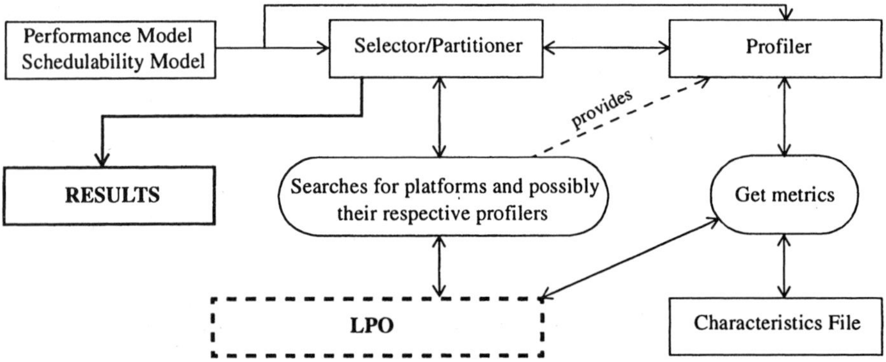

Figure 6-20. Generic usage model for the automated architecture selection and /or system partition algorithms

Architecture selection and system partition algorithms utilize pre-characterized metrics from the platform-independent specification and from the candidate hardware components to determine the feasible target architecture and to partition the specification into hardware and software domains that can be mapped onto the architecture. Such an automated task is very computationally expensive, and asserts no guarantee that the optimal solution will ever be attained, especially where the design space is large and complex. Such shortcomings afflict the usefulness of the automated approach tremendously.

The platform-centric SoC design method specifies a general usage model for the automated architecture selector and/or system partitioner as illustrated in Figure 6.20. In the figure, the selector/partitioner feeds off the performance and/or schedulability model (see UML Real-Time Profile for details [29]) that identifies the desired system objectives for the algorithms. If profiling has to be done on the model, the selector/partitioner either searches the LPO and calls the applicable profiler for each platform or, in a rare case, runs the model against its own profiler to acquire the metrics necessary for the algorithms. Common profiling parameters for candidate platform components, such as throughput or execution time, may be read off the characteristics section of its XML data file, and/or obtained from a dedicated characteristics file maintained elsewhere, possibly as part of the customer support by the selector/partitioner vendors. Once all the required metrics are determined, the selector/partitioner begins its long voyage that usually only ends when certain criteria are met, rather than when the optimal solution is found. Figure 6.21 depicts the performance model for the scenario where the camera is used to take a picture under a Single-Shot, Normal Quality setting. The figure shows how the 1 frame per second requirement, Requirement (NF-R1) in the Supplemental Requirements Document, can be captured. For simplicity, this Sequence model for the JPEG encoding algorithm is not detail-accurate, especially with respect to the IJG implementation.

Table 6-3. Software profiling data on the JPEG algorithm

JPEG Stages	**Percentage of Overall Run Time**					
	Pic. 1	**Pic. 2**	**Pic. 3**	**Pic. 4**	**Pic. 5**	**Pic. 6**
FDCT	77.91	78.85	78.85	77.71	78.12	78.29
Quantize	1.62	1.58	1.55	1.62	1.63	1.58
ZigZag	0.34	0.34	0.34	0.34	0.34	0.34
Huffman Encode	14.01	12.71	12.98	13.9	14.1	13.13

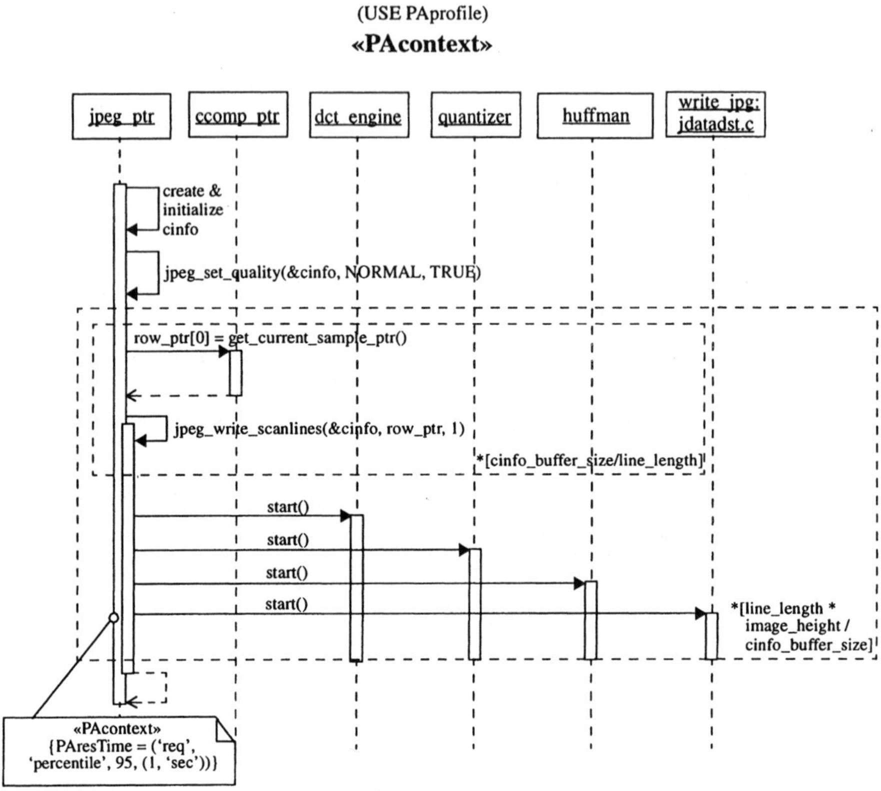

Figure 6-21. Performance model as specified in the UML Real-Time Profile. The figure describes Requirement NF-R1 from the Supplemental Requirements Document

6.4.2 Manual Approach to Selecting Target Architecture

In optimization-based computations found in many different research areas, especially the Operations Research, the initial value plays a crucial role in determining how fast the algorithm converges. In a platform-based co/design problem, the platform represents a pre-determined, and configurable initial values set for a specific problem space. It allows fast convergence or, divergence, if no solution can be found; whereas, the pre-targeted problem space associated with it determines the best-attainable solution for the problem.

For this case study, the inclusion of the Altera's NiOS platform has already been mandated. To get preliminary insight as per how the NiOS system should be configured, the developer may perform a simple profiling on the platform-independent specification, specifically the tasks related to the Take a picture use case for it is where the timing requirement is placed. Table 6.3 shows the JPEG profiling results[2]; while, Chen, et. al. [126] reports similar figures for both Intel PIII 650 MHz, and NiOS 33 MHz (with Software Multiply) platforms. They also present the profiling results of the data handling stage, that consumes approximately 1.72%, and 1.22% of the total execution time (encoding) for the PIII and NiOS platforms, respectively.

It is evident from Table 6.3 and the previous discussion that the FDCT stage is the most computationally expensive. A closer look at the IJG's JPEG library, which is now a part of the platform-independent specification, reveals the support for Fixed- or Floating-point computations, as well as the JPEG algorithms by Arai, Agui & Nakajima (AA&N, see a book by Pennebaker and Mitchell [128] for details), and by Loeffler, Ligtenberg & Moschytz (LL&M) [127]. The AA&N algorithm runs faster (80 multiplies, 464 adds per an 8x8 2D FDCT), but it is less precise when utilizing the Fixed-point computation; whereas the LL&M is more precise, doing 192 multiplies, and 512 adds for the same 8x8 2D FDCT operation. Table 6.4 details possible options for configuring and partitioning the NiOS platform and the platform-independent specification, based on the information acquired from the specification (the JPEG library), and the LPO. Table 6.5 presents the profiles for the NiOS-native multipliers, namely, Software, MSTEP, and MUL, where the MSTEP is a serial multiplier, while the MUL is the parallel implementation.

Table 6-4. System configuration options. By committing to a combination of these options, the developer acquires the target architecture, while simultaneously partitioning the platform-independent specification.

Category	Available Options
Algorithms	AA&N, LL&M
Computational Precision	Fixed-point, Floating-point
Hardware Accelerator	HW Encoder, HW FDCT, HW Multiply, None
Multiply	Software, MSTEP, MUL, Floating-point MUL (implemented using the custom-instruction)
NiOS CPU	16-bit, 32-bit

Table 6-5. Characteristics of the NiOS-native Multipliers

Multiply Type	Logic Elements	Clock cycles (16-bit CPU)	Clock cycles (32-bit CPU)
Software	0	50	250
MSTEP	~20	18	80
MUL	~400	2	16

[2] modified from *http://www.ececs.uc.edu/~ddel/projects/dss/asap/node2.html*

The NiOS system compilation process typically takes about 45-90 minutes on a 256MByte RAM, 600MHz PIII notebook to generate the target system for the developer, making it fairly convenient to compile and re-compile if any of the configurations has to be adjusted. The following decisions, based on the specification and the profiling results, are made for the digital camera system:

- *32-bit CPU*: The 16-bit option might be less effective. The input data are at least 8-bit in size. After a series of additions and multiplications, it is possible that the internal data are going to be larger than 16 bits. Moreover, the 16-bit representation of the internal data might cause the precision error problem to worsen.

- *Hardware Multiply*: It is less costly compared with other hardware accelerators, making it a good starting point. Indeed, a quick look at the hardware encoders as displayed by the POM would unveil that they will not fit in the required EP20K200EFC484-2X PLD device, thus, eliminating them altogether. The MUL option is chosen, to accommodate the multiply-intensive computation of the FDCT (see Table 6.5).

- *Fixed-point*: It is faster than the floating-point counterpart. In addition, there exists no substantial gain in using the floating-point over the fixed-point precision.

- *LL&M*: Because the fixed-point precision is configured, the LL&M algorithm is employed to lessen the effect of the precision error.

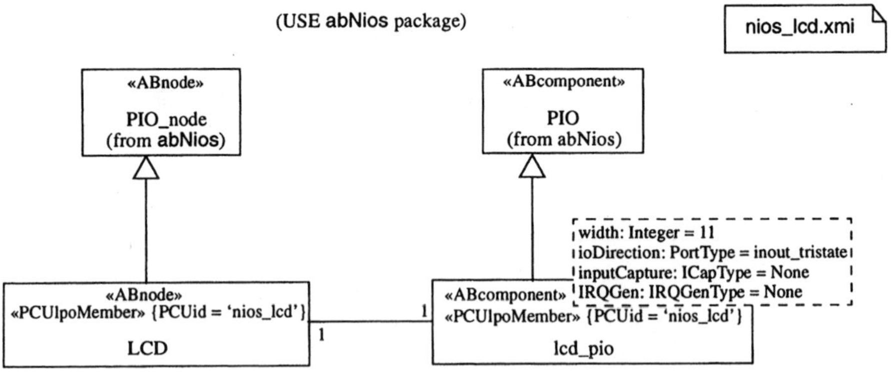

Figure 6-22. UML representation of the LCD. This UML package is accessible through the <uml> tag from within the XML file that describes the LCD (a POmm/component)

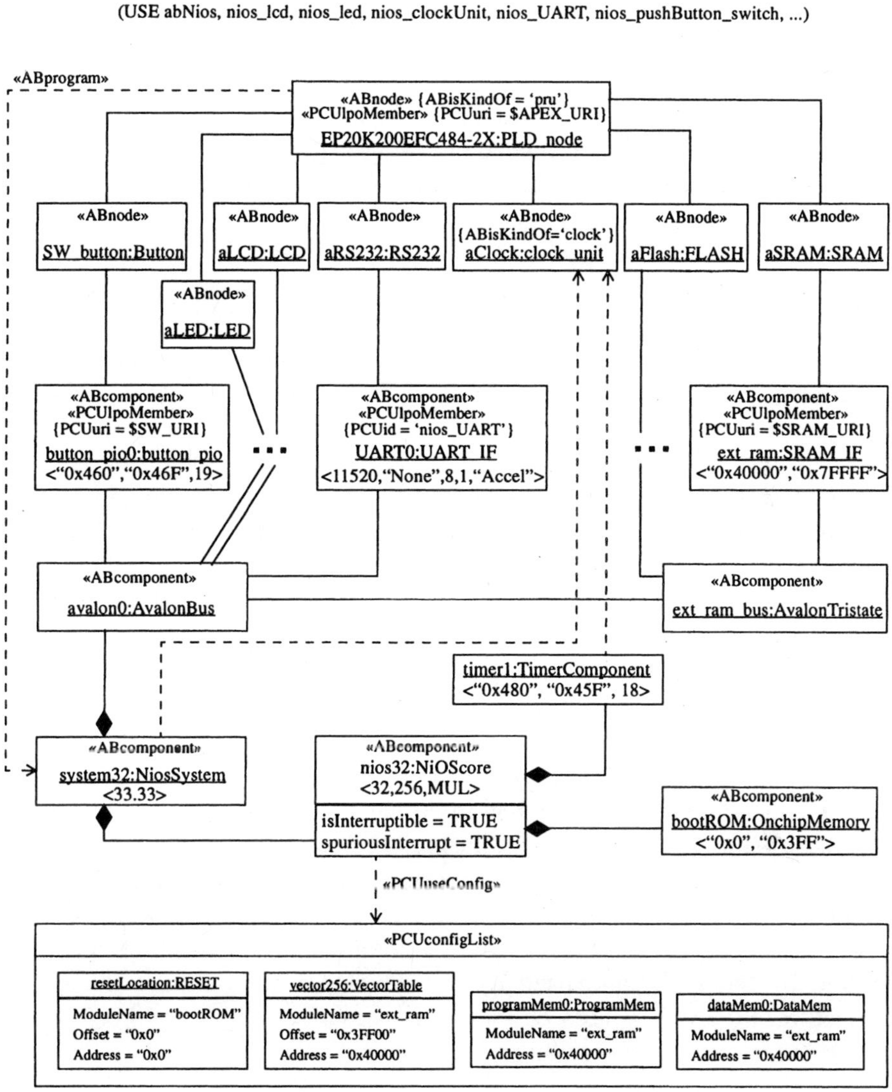

Figure 6-23. The UML description of the candidate target architecture as derived from the blueprint and the associated POmm/components

- *4 Push-button switches* (ID: nios_pushButton_switch), one apiece for the Shutter, Select, Menu, and Done operation.

- *A LED* (ID: nios_LED)

- *RS232/UART* (ID: nios_UART) as a communication means for uploading the images onto the PC.

- Instead of one display as previously specified, a more visible dual seven-segment display unit (ID: nios_dual_7segment_display) is used to show the number of stored images, while a LCD display (ID: nios_LCD) is employed to display text messages.

- 33.33 MHZ clock generator and distributor unit (ID: nios_clockUnit)

- *Flash memory* (ID: Flash_AMD29LVB00) for non-volatile memory, and a SRAM (ID: SRAM_DUAL_IDT71V016) for the executable during prototyping.

It is to note that the decision regarding the peripheral components is made in accordance with their behavioral models described in the platform-independent specification (Figure 6.16). In practice, right-first-time decisions rarely occur, and adjustments as well as fine-grained calibrations are almost unavoidable. The proposed approach helps make the execution of these tedious tasks more tolerable and efficient.

To produce the target architecture, the developer imports the architecture blueprint, and the relevant UML models of the selected peripherals into a common package, and instantiates and configures them as necessary—creating concrete objects from the imported classes. Then, the developer links these objects together in a fashion prescribed by the architecture blueprint, effectively, creating the communications, and completing the instantiation process of the target architecture. Figure 6.22 shows the UML Class diagram representing the NiOS-compliant LCD, which can be retrieved by the POM and used by the developer to configure the target architecture. Figure 6.23 then delineates the resultant target architecture for this particular system configuration.

6.5 Platform-Dependent Specification and System Derivation Process

As opposed to the first two stages, the platform-dependent specification and the system derivation process do not include any iterative sub-process; however, they do traverse back and forth mostly between themselves in an iterative refinement manner that involves (1) the derivation of the platform-dependent specification based on the results from the preceding stages, (2) the activation of the appropriate POmm/tools in the system derivation process whose input is the platform-dependent specification which is being portrayed as different model views in Figure 6.5, and (3) the analysis of the results to further refine the specification, resulting in a loop-back, until all requirements are met. As also indicated in Figure 6.5, if all the requirements could not be attained successfully by refining the platform-dependent specification alone, more fundamental changes may be imperative and the platform-centric process flow leaves the bipartite iteration and goes back to either the platform-independent stage or the platform-analysis stage to make an appropriate modification, before proceeding again into the prescribed development process flow.

6.5.1 Peripheral Interface Routines

In traditional hardware/software codesign approaches, peripheral-interface routines, e.g., device drivers, are normally written by the developer, using low-level macro code specific to the chosen platform. The utilization of the platform-centric SoC design method, or the platform-based approach in general, enables such software routines to be prepared in advance instead by the peripheral and/or platform providers and handed to the developer as a high-level run-time library package that could save the developer time and effort over low-level programming. The exact details as per how such a run-time library package is supported and distributed depends largely on how the platform and the peripheral providers agree on the collaboration framework. For the NiOS platform employed herein, the C run-time library for the relevant peripherals is supported by Altera (the platform provider). The developer accesses the library routines by including the machine-generated nios.h file with the application source code.

The NiOS run-time library is created as a byproduct of the NiOS system generation by the Quartus II software (visit http://www.altera.com for details). At the time of its installation, Quartus II registers itself, i.e. updating the poRegfile, as a POmm/tool that belongs to the NiOS platform. To conform to the platform-centric approach, Quartus II would ideally be able to extract required information right out of the target architecture (Figure 6.23) and produce the run-time libary package as well as the PLD-configuration files, in the SRAM object file (sof) and Intel's hexadecimal output file (hexout) formats. However, to modify this commercial software to be totally compliant with the proposed approach is beyond the means of this book. As a workaround, an intermediate program could be written that analyzes the target architecture in the XMI format, extracts the required information, and generates the Peripherals Template File (ptf), which can then be input to Quartus II. After a successful run, the developer uses the NiOS system library to complete the platform-dependent specification as depicted in Figure 6.25; the report file (rpt) to perform the power estimation

(Figure 6.24); and the sof or hexout file to program the EP20K200EFC484-2X device using the PLD programmer tool listed in Figure 6.17.

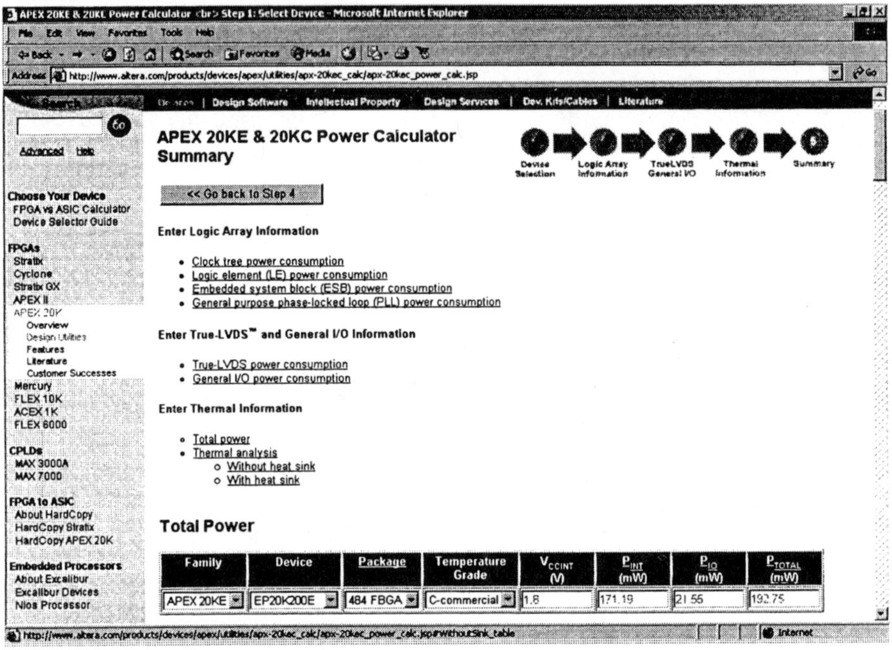

Figure 6-24. The EP20K200EFC484-2X PLD device power calculator provided as a Web application by Altera.

The EP20K200EFC484-2X APEX20KE power calculator, which is supported by Altera and depicted in Figure 6.24, shows the estimated power of 192.75 mW—a fairly reasonable value commercially. Although more optimization effort may ensue to further reduce the power consumption to prolong battery life, this figure already meets the hard requirement that no fan is allowed. The results obtained from this estimator provide a preliminary insight as per the power consumption characteristics of the system under development. More accurate results will manifest later during the test phase.

6.5.2 Transitioning to the Platform-Dependent Specification

As the name implies, the platform-dependent specification infers the structural and behavioral description of a software system to be deployed on a known platform. The developer derives this specification from the platform-independent document and the target architecture model, that are presented in Figures 6.16 and 6.23, respectively.

Table 6-6. The mapping of peripheral-related classes from the platform-independent specification to the platform-dependent specification

Platform-Independent	Platform-Dependent
mssg_service	mssg_service, but it behaves as a unified interface to two new classes, sevenseg_IF and lcd_IF
LED	led_IF
SW_pushButton and its generalized classes	IMprofile facility that specifies the interrupt (button_interrupt), the interrupt service routine (SW_service.start()) and the binding mechanism («IMbind»)
color_components' input buffer (an array representation)	FLASH memory @ 0x104000
media's table of content: - array of image locations - array of image sizes	 FLASH memory @ 0x100000 FLASH memory @ 0x100020
media's effective storage space (an array representation)	FLASH memory @ 0x120000

Table 6.6 shows the mapping of the peripheral-related classes from the platform-independent specification to its platform-specific counterpart. As evident from the table, the mapping is fairly minimal and systematic, where the peripheral-related classes are replaced by their respective interface classes and the array pointers are adjusted to reflect the chosen addresses[3]. Such relative ease can be attributed to their explicit modeling in the platform-independent specification that permits them to be treated as objects, and become more structural and reusable (see also Section 6.3.3). To emphasize the point, consider the UART which is never explicitly modeled. Without the benefit of class reuse, the developer would have to modify every single occurrence of the UART-related operations—a process which can be tedious and error-prone. In this case, the UART is only used within the upload_method class, and involves only the overloaded printf() function that can output the printed string to the GERMS monitor via the RS232 PC serial port. Figure 6.25 depicts the detail-suppressed class model of the platform-dependent specification.

[3] The NIOS system utilizes a memory-mapped architecture

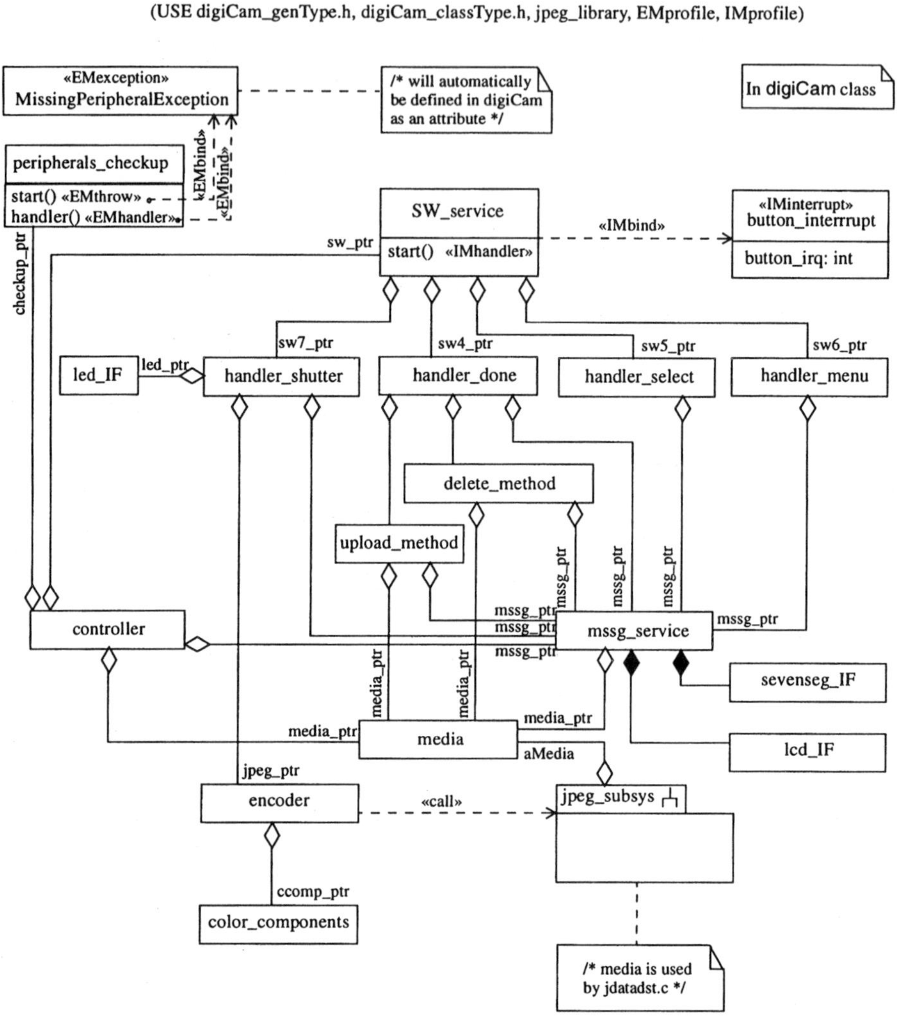

Figure 6-25. Detail-minimal platform-specific Class diagram describing the digital camera system

6.5.3 Deriving the System

The sheer complexity of today's SoC systems development mandates that automated tools be an integral part of the design process. The presence of such tools in a well-integrated environment can enhance the prescient insight of the developer, resulting in an effective decision making process that can expedite the system development as a whole.

As previously mentioned, the system derivation process is an iterative process involving different tools for different model views that represents different aspects of the system under development. To further drive the refinement process of the digital camera system development, the developer can re-target the profiling of the JPEG compression task against the preliminary system architecture so as to expose additional expectable, architecture-dependent system characteristics. Table 6.7 tabulates the execution times of the JPEG encoder, along with its FDCT sub-module's, that result from the profiling against the compression quality values of 65, 75, 90 and 99. The profiling input is a 227x149 RGB raw color components of size 33.03 KBytes, and the target architecture is the one illustrated in Figure 6.23.

By perceptive inspection, no drastic difference among the four images is found. Consequently the developer specifies the compression quality values, and makes the requirements adjustment. The NORMAL image quality is therefore defined to correspond to the compression quality of 65, and 90 is defined for the GOOD image quality.

A simple linear projection can be performed that could roughly estimate the worst-case execution time for the NORMAL and GOOD image compression on the required 640x480 RGB input components (Requirement NF-R6). By using the input sample size of 14400 (640x480x3) with the timing characteristics as shown in Table 6.8 that come directly from Table 6.7, the developer obtains the estimated worst-case execution times of 9.216 and 11.232 seconds for the NORMAL and GOOD setting, respectively. Both of these profiling results are greater than the 1 second requirement. Hence, the developer performs additional profiling on different settings and architectures, the results of which are shown in Table 6.8. Notice that the hardware encoder option is not included because it does not fit into the EP20K200EFC484-2X device, whose useable resource is 8320 LEs, while all of the encoder implementations (from Amphion, Barco Silex, and CAST) require the LEs in excess of this threshold.

Table 6-7. Profiling results on timing characteristics of the JPEG compression (LL&M algorithm) of the 227x149 RGB color input components with respect to different compression quality values

Compression Quality	**Encoder**	**FDCT (851 blocks)**	**Non-FDCT**	**JPEG Output Size**
65	861.53 msec	203.97 msec	657.56 msec	4.75 KBytes
75	885.9 msec	204 msec	681.9 msec	5.02 KBytes
90	1095 msec	203.98 msec	891.02 msec	10.0 KBytes
99	2037 msec	203.98 msec	1833.02 msec	27.7 KBytes

Table 6-8. Profiling results on timing characteristics of the JPEG compression of the required 640x480 RGB color input components subject to different configurations

Timing Characteristics (msec per 2D 8x8 block) Non-FDCT (NORMAL): 0 40, non-FDCT (GOOD): 0.54, DCT (AA&N): 0.19, DCT (LL&M): 0.24					
Algorithms	**HW Accelerators**	**Settings**	**Total (seconds)**	**Non-FDCT (seconds)**	**FDCT (seconds)**
LL&M	MUL	NORMAL	9.216	5.760	3.454
		GOOD	11.232	7.776	3.456
	HW FDCT (with DMA)	NORMAL	5.760	5.760	Negligible
		GOOD	7.776	7.776	Negligible
AA&N	MUL	NORMAL	8.496	5.760	2.736
		GOOD	10.512	7.776	2.736
	HW FDCT (with DMA)	NORMAL	5.760	5.760	Negligible
		GOOD	7.776	7.776	Negligible

The profiling results in Table 6.8 suggest that (1) there exists no feasible solution to the problem, i.e. given the current requirements specification and the availability of the POmm/components, the developer cannot achieve the targeted digital camera system without the specification violation, and (2) a constraint relaxation is needed that could result in the modification of the requirements specification. Indeed it is quite ambitious to implement a JPEG encoder on a 33 MHz, moderate density PLD device that can compress a 640x480 RGB image in 1 second. The following detail some plausible scenarios assuming that the decision is made in favor of relaxing the constraints.

- Another PLD device is acquired that is more powerful, such that it can house a complete hardware implementation of the JPEG encoder,

- The requirement reduces the number of input samples, and effectively, leaves the competitive digital camera market (maybe, entering the cellular phone market instead?). This decision would also affect the criteria for choosing an image sensor for the digital camera system.

By employing a more powerful, higher density PLD device, the developer could eradicate the bottleneck caused by slow clock speed, and high-volume data transfer. The pursuit of this option would likely yield the feasible digital camera system. However, it would incur additional cost of improving a PLD device, and developing/acquiring the hardware JPEG encoder.

On the other hand the developer could compute a more suitable input image size based on the data in Table 6.8. Because of the explicit 1 fps requirement on the NORMAL setting operation, and also because of the use of the fixed-point arithmetic for the JPEG engine, the data from the LL&M, NORMAL settings are used in the calculation. Through a simple linear interpolation of the digital camera performance where the system with the FDCT hardware accelerator runs at 0.4 msec/block, and that with the hardware multiply (MUL) at 0.64 msec/block, the developer attains the estimated input sample size of 53312, and 33344 for the hardware FDCT and MUL multiply, respectively. These numbers can be translated into the image dimension of approximately 168x320, and 168x200 pixels, both of which are much smaller than the 640x480 tentative specification.

6.5.4 Concluding Remarks

Thus far, this book has presented, in a somewhat pedagogical manner, the development process for the simplified digital camera system based on the proposed platform-centric SoC design method. It has demonstrated the strength of the proposed approach as an analysis, modeling, design, and documentation means that could expedite the complex SoC system development process via a well-integrated design environment that embraces the exploitation of platforms, UML and XML technologies.

Where the task description ended in the previous section, the developer could have resumed by deciding on the available options and proceeded until the final system would result that meets all the specified requirements. Since all the steps involved have already been discussed, the book halts here to avoid redundancy. The next section presents the implementation results, and compares the cost-effectiveness of the proposed approach against that of the SpecC methodology.

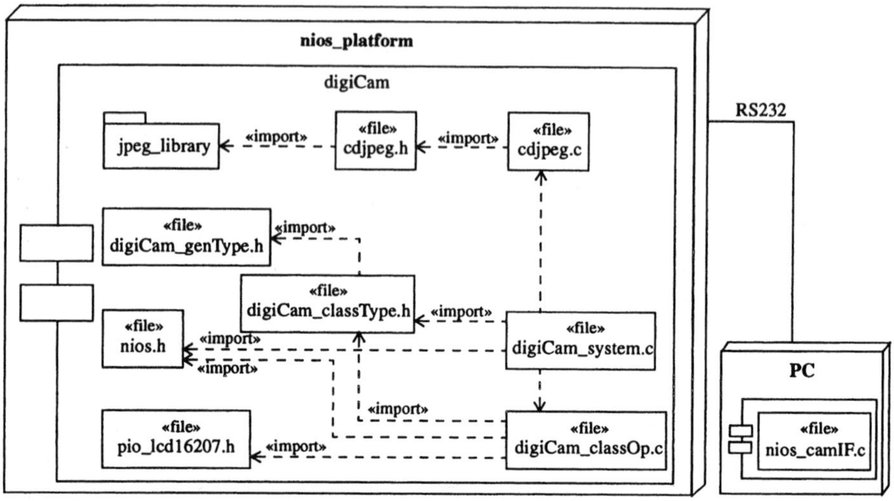

Figure 6-26. The source files hierarchy for the digital camera system

6.6 Implementation Results

The eventual implementation of this digital camera system conforms to the platform-dependent specification as discussed in Section 6.4 (see also Figures 6.23, and 6.25), with the exception that the input image dimension has effectively been scaled down to 160x160, instead of 640x480 as required by the early specification (see Section 6.5.3 for details). Figure 6.26 depicts the source files hierarchy of the implementation.

6.6.1 Resultant Timing and Compression Characteristics

C is one of the most popular programming languages currently used to develop real-time embedded applications. To demonstrate that C and UML can work well together despite their object-oriented discrepancy, all source files are presently implemented in C in a form that mimics the look and feel of an OO programming language. Such a structural approach incurs an estimate 85% increase in code size compared with the traditional approach, but produces a much better correlation between the UML model and its corresponding C code.

Table 6.9 tabulates the timing and compression characteristics for different input images, all of which are represented by the 160x160 RGB components of size 25KBytes. As evident from the results, this fully-functional digital camera system prototype performs reasonably well, exhibiting only a single timing violation (sample06.jpg), while all natural scenes yield the timing results under the 1 second mark. Further extensive investigation will determine if this system actually fulfills the 95%, 1 fps requirement as dictated in the specification.

Table 6-9. Timing and compression characteristics data from different input images

Input Image Characteristics: 160x160 RGB, 25 KBytes				
Input Images	**Settings**	**Total (seconds)**	**JPEG file size (KBytes)**	**% Compressed**
 sample01	NORMAL GOOD	0.515 0.922	2.70 8.44	89.2 66.24
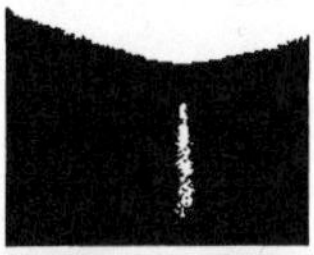 sample02	NORMAL GOOD	0.523 0.957	2.99 9.83	88.04 60.68
 sample03	NORMAL GOOD	0.545 0.987	3.78 10.9	84.88 56.4
 sample04	NORMAL GOOD	0.722 1.213	5.59 14.8	77.64 40.8
 sample05	NORMAL GOOD	0.948 1.678	9.18 23.1	63.28 7.6
 sample06	NORMAL GOOD	1.028 1.841	11.5 24.7	54 1.2
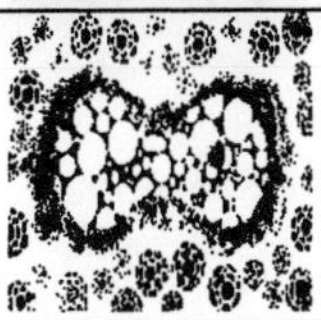 sample07	NORMAL GOOD	0.743 1.457	6.45 19	74.2 24

6.6.2 Research Evaluation

This section introduces the notion of cost as a metric for evaluating the robustness of the proposed approach. The cost modeling technique estimates the system development cost by taking into account a number of factors, from human to technology to the complexity of the project itself, that affect the development process. Such a technique could be very accurate when carefully calibrated to a specific problem, yet it normally yields good results, even without much calibration, when used to compare cost-effectiveness of two or more development costs. In the discussion to follow, the COCOMO II.2000 [19] cost modeling technique is utilized that comparatively evaluates the cost-effectiveness of the proposed approach against the SpecC methodology. At present SpecC is undergoing a standardization process to become a standard language and interchange format for system specification design (see http://www.specc.org). Thus, it constitutes an eligible benchmark for an evaluation of this research.

Typically, a development cost of an embedded system equals a sum of costs incurred by software and hardware, plus any cost or revenue adjustment amounted from missing or meeting the window of opportunity when the product enters the market. For an evaluation of this research, however, only the software cost is used in the calculation—a decision attributable to the fact that the SpecC and the proposed approaches permit the developer to work at too high a level of design abstraction for the current hardware cost modeling technique to be an effective efficiency indicator. On the other hand, the window of opportunity adjustment, which is computed by using the time-to-market cost estimation model [6, 21, 22], is directly proportional to the results from the software cost estimation, and hence, can be ignored when comparing cost-effectiveness of the two approaches.

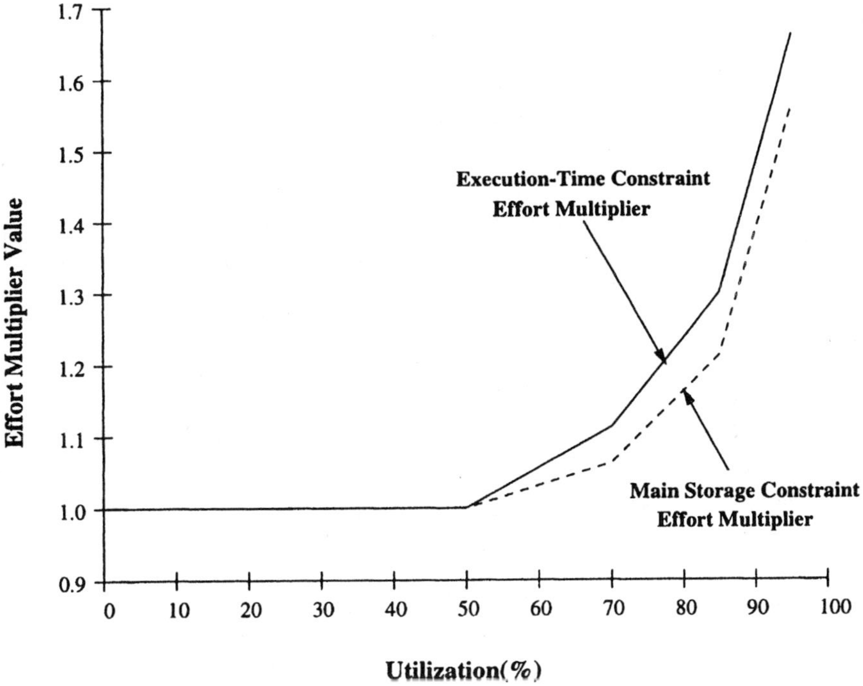

Figure 6-27. Execution time and main storage constraint effort multipliers vs resource utilization

6.6.2.1 Software Cost Modeling

Software cost estimators primarily consist of a core or nominal effort equation that relates the labor effort for developing software to the size of the software system. This nominal effort equation represents the cost of developing a software system under ideal conditions. To get a more realistic view of such software cost, effort adjustment factors are applied to the nominal estimate to adjust for organization and project-specific economic factors. For example, the COCOMO II [19] model uses the adjustment factors as shown in Table 6.10 to make the nominal estimate more realistic.

Although many of these tools employs very different parametric cost estimating relationships (CERs), they all make similar claims about how the design can affect the software development cost. In all cases, historical data show that increased development cost and time can occur as a result of squeezing more and more functionality in smaller and smaller space and time intervals, incurring more design efforts and overall cost as a result.

REVIC [12], COCOMO II [19], and PRICE-S [18], for example, assume that resource requirements of less than 50% capacity have no cost impact, but as the utilization approaches 100% the cost impact becomes extreme as depicted in Figure 6.27 [16].

6.2.2.2 Cost Comparison

The COCOMO II.2000 parametric cost model [19] is used to evaluate the robustness of this research against the SpecC methodology based on cost-effectiveness of the two approaches. The COCOMO II.2000 model estimates the system development effort using the equation of the form:

$$C_E = A(KSLOC)^E \prod_{i}^{1} EM_i \qquad (6.1)$$

where the unit of the effort equation (C_E) is in person-months, i.e. the amount of time one person spends working on the system development project for one month. The coefficient A is a productivity constant which captures the effects on effort with projects of increasing size. E, an aggregation of five scale factors (SF), accounts for the relative economies of scale (values less than 1.0) or diseconomies of scale (values greater than 1.0) encountered for projects of different sizes. Its value can be computed according to the equation below:

$$E = B + 0.01 \sum_{j}^{5} SF_j \qquad (6.2)$$

where the constant B = 0.91.

The effort multipliers EM_i model the effect of personnel, computer, product, and project attributes on software development cost. Table 6.10 gives a brief description of the COCOMO II.2000 effort multipliers.

Table 6-10. COCOMO II.2000 effort multipliers (EMs).

Cost Drivers	Ratings
Product Factors	
EM1: Required software reliability (RELY)	0.82 to 1.26
EM2: Database size (DATA)	0.90 to 1.28
EM3: Product complexity (CPLX)	0.73 to 1.74
EM4: Developed for reusability (RUSE)	0.95 to 1.24
EM5: Documentation match to life-cycle needs (DOCU)	0.81 to 1.23
Platform Factors	
EM6: Execution-time constraint (TIME)	1.00 to 1.63
EM7: Main-storage constraint (STOR)	1.00 to 1.46
EM8: Platform volatility (PVOL)	1.00 to 1.30
Personnel Factors	
EM9: Analyst capability (ACAP)	0.71 to 1.42
EM10: Programmer capability (PCAP)	0.76 to 1.34
EM11: Personnel continuity (PCON)	0.81 to 1.29
EM12: Applications experience (APEX)	0.81 to 1.22
EM13: Platform experience (PLEX)	0.85 to 1.19
EM14: Language and tool experience (LTEX)	0.84 to 1.20
Project Factors	
EM15: Use of software tools (TOOL)	0.78 to 1.17
EM16: Multisite development (SITE)	0.80 to 1.22
EM17: Required development schedule (SCED)	1.00 to 1.43

However, the primary input to the COCOMO II.2000 cost estimator is the software size estimate, *KSLOC*. *KSLOC* denotes the number of source lines of code (thousands), which include application code, OS kernel services, control and diagnostics, and support software. Software size estimates comprise two parts: the number of new source lines of code (*KNSLOC*), and the number of adapted source lines of code (*KASLOC*)—both are represented as a numerical multiplication of one thousand. This evaluation uses the following COCOMO II.2000 model to calculate the total number of source lines of code:

$$KSLOC = KNSLOC + (KRSLOC \times AAM_R) + (KASLOC \times AAM_A) \quad (6.3)$$

where AAM_R and AAM_A has the general form:

$$AAM = .1(AA + ((.4(DM) + .3(CM) + .3(IM)) \times (1 + .02(SU)(UNFM))) \quad (6.4)$$

The symbols in the equations are defined in Table 6.11.

Table 6-11. Software Sizing Model Symbol Definitions

Symbols	Description
KNSLOC	Size of component expressed in thousands of new source lines of code
KRSLOC	Size of the reused software component expressed in thousands of adapted source lines of code
KASLOC	Size of the adapted software component expressed in thousands of

	adapted source lines of code
AAM	Adaptation adjustment modifier
AA	Degree of assessment and assimilation
DM	Percentage of design modified
CM	Percentage of code modified
IM	Percentage of integration and test modified
SU	Software understanding penalty
UNFM	Software unfamiliarity

Before determining the number of effective source lines of code (*KSLOC*) resulted from employing the SpecC approach to develop the specified digital camera system, this book recounts the principal tasks as prescribed by Figure 6.28 as follows (see also Section 2.3):

- *Specification phase*. SpecC derives the system specification during this phase of operation. As opposed to the proposed approach, it specifies no analysis process as per how the specification might be derived from the initial requirements. The resultant specification comprises the JPEG encoder, which can be obtained from *http://www.ics.uci.edu/~specc/*, as well as the behavioral specification of the digital camera interface.

- Architecture phase. As shown in Figure 6.28, this phase involves the architecture selection task and the system partition task. SpecC exploits its own profiling tool to extract the metrics necessary to perform these tasks from the NiOS-based hardware components, assumed to be available for the developer. At the end of this phase, a more refined specification results that is an equivalent of the platform-dependent specification in the platform-centric SoC design method.

- Communication phase. During this phase, the communication channels between hardware and software are modeled, elaborated and implemented. The result is a refined architecture model with all the abstract communication channels being resolved and synthesized. At the end of this phase, SpecC produces a component netlist in VHDL and software I/O instructions in C as its output

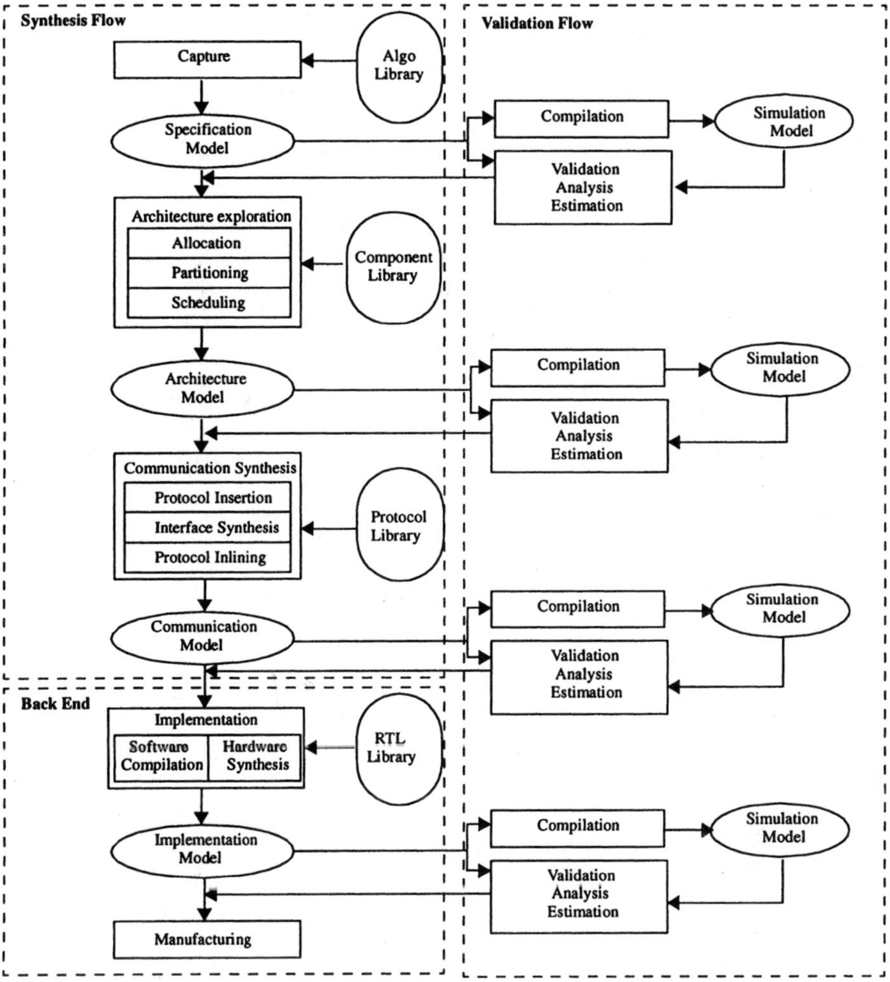

Figure 6-28. The SpecC methodology process flow

- Implementation phase. The developer uses the Cygwin C/C++ compiler to compile the C code. The hardware portion of the model, including the interfaces between hardware and software, is synthesized using Leonardo Spectrum. This design phase generates the implementation model which consists of object codes executing on the NiOS processor and a gate-level netlist of the hardware components which must then be fed to Quartus II for I/O pin connections and the *sof* file generation.

As identifiable from the SpecC tasks description above, the non-trivial source code that contributes to the *KSLOC* count come from the SpecC

JPEG encoder specification model, the camera interface's behavioral description, and the NiOS specific software I/O instructions[4]. Because SpecC is an OO superset of C, the source lines for the interface behavioral description is estimated to be equal to the traditional C implementation of the same specification, which was written for the platform-centric development process for fast verification purpose. On the other hand, the lines count for the NiOS specific software I/O instructions is taken directly form the Quartus II generated run-time library. Table 6.12 summarizes the relevant numbers of source lines of code for both approaches.

Derived from Table 6.12, and based on Equations 6.3 and 6.4 , the effective source lines of code for the SpecC, and the proposed approach are 2.46, and 1.13 *KSLOC*. Then, by applying Equations 6.1 and 6.2, the estimated incurring costs for the development of the digital camera system using the SpecC and the platform-centric SoC design method are 5.0, and 1.6 person-months, respectively.

6.6.3 Concluding Remarks

Cost-wise, the proposed approach appears to benefit greatly from the use of a platform, and the use of a pre-existing IJG's JPEG library, contributing to a smaller *KSLOC* value, regardless of the total number of source lines which is much larger than that of the SpecC (9.202K to 3.353K). Such results serve to further attest the role of reuse as indeed being indispensable in all phases of the complex system on chip (SoC) development process.

Table 6-12. Summary of the source lines of code applicable to the SpecC and platform-centric approaches

Relevant *KSLOC*	**SpecC**		**Platform-Centric**	
	Size (thousand)	**Type**	**Size (thousand)**	**Type**
Structural C implementation of digiCam	n/a	n/a	0.936	*KNSLOC*
Traditional C implementation of digiCam[5]	0.567	*KNSLOC*	n/a	n/a
jdatadst.c	n/a	n/a	0.121	*KASLOC*
IJG JPEG library	n/a	n/a	6.345	*KRSLOC*
SpecC JPEG encoder[4]	0.986	*KASLOC*	n/a	n/a

[4] The hardware glue logic is assumed to be small.

[5] Represents the *best-case* estimate of the actual SpecC implementation.

Software I/O routines	1.8	*KNSLOC*	1.8	*KRSLOC*

Chapter 7

SUMMARY

7. CONTRIBUTIONS

This book has proposed, developed and implemented a platform-centric SoC design methodology, based on the Co-design Modeling Framework (CMF) profile of UML, that promises improvements in cost and technology-to-market time for the development of complex systems, while also effectively enhancing design space exploration.

The key highlights of this book include:

- Enabling technologies: Platforms, UML, and XML, were identified, and seamlessly integrated that contribute to the development of a flexible and robust system design process.

- A reuse-based design methodology: The proposed platform-centric SoC design methodology fosters reuse of both UML models (abstract platform components) and non-UML models (e.g., IJG's JPEG library), as well as the use and reuse of knowledge through the WWW technologies. In addition, it allows the internal processes to vary that could be more fitting to the chosen platform.

- Integration with UML and XML: A unified, visual representation of the system under development was developed as the UML profile for Codesign Modeling Framework (CMF), based on UML 1.5 [24] and the UML Real-Time profile [29], that can be utilized within the proposed platform-centric environment for modeling, design, analysis, synthesis, implementation and documentation purposes. The CMF profile allows all aspects of the SoC system development process, right from the initial requirements, to be described using one common language for better efficiency.

- LPO Library Specification: The specification of an XML database, known as the Library of Platform Objects (LPO) is described in detail, along with its anticipated usage and behaviors. The LPO could span the whole Internet space, and could be distributed—paving a way for interoperability standards and a pathway for the system design community to converge on a few standard platforms. The task of populating platform components and tools could be performed in a standard way, not limited to any one individual or organization, and these components and tools could enter and exit the LPO freely, so long as they are Internet-accessible.

- Application of UML: The use of UML to assist in the development of a complex hardware/software co-design system, such as a digital camera, that involves real-time characteristics was demonstrated. It was also shown that such a UML application efficiently empowered an incorporation of OO analysis techniques, as well as enhanced design reuse, resulting in an overall improvement of the platform-centric SoC design approach.

7.1 Future Directions – A Multi-Disciplinary Field

This book is the product of integrating technologies in three widely different disciplines, namely platforms, the UML, and XML, so as to form a robust system development approach that is applicable for use with a multitude of complex system requirements today and tomorrow. Each one of these technologies has gained steady research interest in both the industry and academia, and would likely continue to contribute to the proposed platform-centric approach. Nonetheless, to reap full benefits proffered by the approach, there needs to be agreement reached between all involved (and vested) parties—the system houses, IC manufacturers, and tools and components providers.

This book envisages "reuse" as the principal driving force that contributes to an efficient system development process. A tool-integrated environment for the POM represents another intriguing research area within this novel approach. Since the POM relates closely with XML and the Internet technologies, how such research disciplines as Network

Programming and Distributed Computing can enhance the POM capability would be worth an investigation. A UML-capable, UML-RT/CMF enabled POM would also be ideal for the approach.

In addition, this book assumes that platforms are pre-built, and readily available to the system developer. The actual development of a platform, however, is an extremely complex process as evidenced by pioneers, such as Karl Sabbagh [96], Alberto Sangiovanni-Vincentelli [14] and other related work [70, 129]. More research is needed in this area to reduce the platform development cost, while efficiently producing a system platform that is well-suited to as many applications as possible.

REFERENCES

1] J. Turley, "The Two Percent Solution," *Embedded.com*, December 2002; http://www.embedded.com/story/OEG20021217S0039.

[2] C. Herring, "Microprocessors, Microcontrollers, and Systems in the New Millenium," *IEEE Micro*, vol. 20, no. 6, pp. 45-51, November-December 2000.

[3] V. J. Mooney III, D. M. Blough, "A Hardware-Software Real-Time Operating System Framework for SoCs," *IEEE Design & Test of Computers*, vol. 19, no. 6, pp. 44-51, November-December 2002.

[4] R. Mahajan, K. Brown, V. Atluri, "The Evolution of Microprocessor Packaging," *Intel Technology Journal*, Intel Corp., 2000; http://developer.intel.com/technology/itj/q32000/articles/art_1.htm.

[5] L. F. Friedrich, J. Stankovic, M. Humphrey, M. Marley, J. Haskins Jr., "A Survey of Configurable, Component-Based Operating Systems for Embedded Applications," *IEEE Micro*, vol. 21, no. 3, pp. 54-67, May-June 2001.

[6] J. A. DeBardelaben, "An Optimization-Based Approach for Cost-Effective Embedded DSP System Design," *Ph.D. Thesis, Georgia Institute of Technology*, May 1998.

[7] J. A. DeBardelaben, V. K. Madisetti, A. Gadient, "Incorporating Cost Modelign in Embedded-System Design," *IEEE Design & Test of Computers*, pp. 24-35, July-September 1997.

[8] B. C. Cole, "Getting to the Market On Time," *Electronics*, pp. 62-65, April 1989.

[9] D&T Roundtable, "Hardware-Software Codesign," *IEEE Design & Test of Computers*, vol. 17, no. 1, pp. 92-99, January-March 2000.

[10] F. Slomka, M. Dorfel, R. Munzenberger, R. Hofmann, "Hardware/Software Codesign and Rapid Prototyping of Embedded Systems," *IEEE Design & Test of Computers*, vol. 17, no. 2, pp. 28-38, April-June 2000.

[11] University of California, Berkeley, A Framework for Hardware-Software Co-Design of Embedded Systems, POLIS Release 0.4, December 1999; http://www-cad.eecs.berkeley.edu/Respep/Research/hsc/abstract.html.

[12] U.S. Air Force Analysis Agency, *REVIC Software Cost Estimating Model User's Manual Version 9.2*, December 1994.

[13] M. Sgroi, L. Lavagno, A. Sangiovanni-Vincentelli, "Formal Models for Embedded System Design," *IEEE Design & Test of Computers*, vol. 17, no. 2, pp. 14-27, April-June 2000.

[14] A. Sangiovanni-Vincentelli, G. Martin, "Platform-Based Design and Software Design Methodology for Embedded Systems," *IEEE Design & Test of Computers*, vol. 18. no. 6, pp. 23-33, November-December 2001.

[15] Department of Defense, *Parametric Cost Estimating Handbook*, Fall 1995.

[16] B. Boehm, *Software Engineering Economics*, Prentice-Hall, Inc., Englewood Cliffs, NJ, 1981.

[17] D. Lyons, "SEER Helps Keep Satellite Cost Estimates Down to Earth," *Infoworld*, November 27, 1995.

[18] A. Minkiewicz, A. DeMarco, "The PRICE Software Model," *PRICE Systems Intenal Document*, June 1995.

[19] University of Southern California, Center for Software Engineering, *COCOMO II: Model Definition Manual*, Version 2.1, 2000.

[20] D. Gajski, J. Zhu, R. Dömer, A. Gerstlauer, S. Zhao, *SpecC: Specification Language and Design Methodology*, Kluwer Academic Publishers, March 2000.

[22] J. Miranda, "A BIST and Boundary-Scan Economics Framework," *IEEE Design & Test of Computers*, pp. 17-23, September 1997.

[23] J. Axelsson, "Analysis and Synthesis of Heterogeneous Real-Time Systems," *Ph.D. Thesis, Linköping University*, 1997.

[24] The Object Management Group, *OMG Unified Modeling Language Specification Version 1.5*, March 2003; http://www.omg.org/.

[25] M. Fowler, *UML Distilled: A Brief Guide to the Standard Object Modeling Language*, 2nd Edition, Addison-Wesley, 2000.

[26] B. Mathew, "UML Tutorial," *Stylus Inc.'s Online Articles*, Accessed: January 2003; http://www.stylusinc.net/articles/uml/lesson1.shtml.

[27] "Practical UML: A Hands-On Introduction for Developers," *TogetherSoft Inc.*; http://www.togethersoft.com/services/practical_guides/umlonlinecourse; accessed January 2003.

[28] B.P. Douglass, *Doing Hard Time: Developing Real-Time Systems with UML, Objects, Frameworks, and Patterns*, Addison-Wesley, 1999.

[29] The Object Management Group (OMG), UML Profile for Schedulability, Performance, and Time Specification, March 2002; http://www.omg.org/technology/documents/modeling_spec_catalog.htm.

[30] G. Booch, J. Rumbaugh, I. Jacobson, *The Unified Modeling Language User Guide*, Addison-Wesley, 1999.

[31] "NRE Charges for SoC Designs to Hit $1M, Study Says," *Semiconductor Business News*, October 1999; http://www.csdmag.com/story/OEG19991006S0043.

[32] S. P. Dean, "ASICs and Foundries," *Integrated Systems Design*, March 2001; http://www.isdmag.com/story/OEG20010301S0051.

[33] *DOmain Modeling Environment, Version 5.3*, Honeywell Labs, Morristown, NJ, August 2000; http://www.htc.honeywell.com:80/dome/description.htm.

[34] *ArgoUML Community Version, Version 0.12*, BSD License, January 2003; http://argouml.tigris.org/.

[35] G. Kotonya, I. Sommerville, *Requirements Engineering: Processes and Techniques*, John Wiley & Sons, 1998.

[36] L. A. Maciaszek, *Requirements Analysis and System Design: Developing Information Systems with UML*, Addison Wesley, 2001.

[37] M. Edwards, "Software Acceleration Using Coprocessors: Is It Worth the Effort?," *Proceedings of the 5th International Workshop on Hardware/Software Codesign*, pp. 135-139, 1997.

[38] G. M. Amdahl, "Validity of Single-Processor Approach to Achieve Large-scale Computing Capability," *Proceedings of the 30th AFIPS Spring Joint Computer Conference*, pp. 483-485, 1967.

[39] R. Ernst, J. Henkel, T. Benner, "Hardware-Software CoSynthesis for Microcontrollers," *IEEE Design & Test of Computers*, vol. 10, no. 4, pp. 64-75, December 1993.

[40] R. Gupta, G. De Micheli, "Hardware-Software CoSynthesis for Digital Systems," *IEEE Design & Test of Computers*, pp. 29-41, September 1993.

[41] T.-Y. Yen, W. Wolf, "Sensitivity-Driven Co-synthesis of Distributed Embedded Systems," *Proceedings of the 8th International Symposium on System Synthesis*, 1995.

[42] T.-Y. Yen, W. Wolf, "Communication Synthesis for Distributed Embedded Systems," *Proceedings of the IEEE International Conference on Computer-Aided Design*, pp. 288-294, 1995.

[43] S. Schulz, J.W. Rozenblit, K. Buchenrieder, "Towards an Application of Model-Based Codesign: An Autonomous, Intelligent Cruise Controller," *Proceedings of the 1997 IEEE Conference and Workshop on Engineering of Computer Based Systems*, Monterey, CA, pp. 73-80, March 1997.

[44] K. Buchenrieder, J.W. Rozenblit, "Codesign: An Overview," *Codesign: Computer-Aided Software/Hardware Engineering*, pp. 1-16, IEEE Press, 1994.

[45] University of California, Berkeley, *The Ptolemy Project*, accessed February 2003; http://ptolemy.berkeley.edu/.

[46] R. Goering, "24-hour Chip Design Cycle Called Possible," *EETIMES*, August 2001; http://www.eedesign.com/story/OEG20010803S0083.

[47] S. Holzner, *Inside XML*, New Riders Publishing, pp. 1-14, November 2001.

[48] A. Møller, M. I. Schwartzbach, "The XML Revolution: Technologies for the Future Web," *BRICS, University of Arthus*, October 2002: http://www.brics.dk/~amoeller/XML/.

[49] A. Bergholz, "Extending Your Markup: An XML Tutorial," *IEEE Internet Computing*, July-August 2000; http://computer.org/internet/xml/xml.tutorial.pdf.

[50] J. P. Thomas, "Understanding XML (eXtensible Markup Language)," *Stylus Inc.'s Online Articles*; http://www.stylusinc.net/technology/XML/xml_future.shtml; accessed March 2003.

[51] M. Bryan, "An Introduction to the Extensible Markup Language (XML)," http://www.personal.u-net.com/~sgml/xmlintro.htm; accessed March 2003.

[52] *Document Object Model (DOM),* World Wide Web Consortium; http://www.w3.org/DOM/, accessed March 2003.

[53] *The Common Object Request Broker: Architecture and Specification Version 3.0.2*, The Object Management Group (OMG); http://www.omg.org/technology/documents/corba_spec_catalog.htm, accessed March 2003.

[54] N. Walsh, "A Technical Introduction to XML," *World Wide Web Journal*, Revision 1.1, February 1998; http://www.w3j.com; http://nwalsh.com/docs/articles/xml/index.html.

[55] T. Bray, J. Paoli, C. M. Sperberg-McQueen, E. Maler, *Extensible Markup Language (XML) 1.0, Second Edition*, World Wide Web Consortium, October 2000; http://www.w3.org/TR/REC-xml.html.

[56] *XML Schema 1.0*, World Wide Web Consortium, May 2001; http://www.w3.org/XML/Schema.

[57] OASIS, "Extensible Markup Language (XML)," *Cover Pages*, February 2003; http://xml.coverpages.org/xml.html.

[58] S. DeRose, E. Maler, D. Orchard, *XML Linking Language (XLink) Version 1.0*, World Wide Web Consortium, June 2001; http://www.w3.org/TR/xlink/.

[59] S. DeRose, R. Daniel, et.al., *XML Pointer Language (XPointer),* World Wide Web Consortium, August 2002; http://www.w3.org/TR/xptr/.

[60] J. Clark, S. DeRose, *XML Path Language (XPath) Version 1.0*, World Wide Web Consortium, November 1999; http://www.w3.org/TR/xpath.

[61] *Resource Description Framework (RDF),* World Wide Web Consortium, February 1999; http://www.w3.org/RDF/.

[62] F. Manola, E.Miller, et.al., *RDF Primer*, World Wide Web Consortium, January 2003; http://www.w3.org/TR/rdf-primer/.

[63] S. Bradner, "Key words for use in RFCs to Indicate Requirement Levels (RFC2119)," *The Internet Engineering Task Force*, March 1997; http://www.ietf.org/rfc/rfc2119.txt.

[64] C. Alexander, S. Ishikawa, M. Silverstein, *A Pattern Language: Towns, Buildings, Construction*, Oxford University Press, 1977.

[65] C. Alexander, *The Timeless Way of Building*, Oxford University Press, 1979.

[66] B. T. Kurotsuchi, "Design Patterns," *Online Tutorial*, 1996; http://www.csc.calpoly.edu/~dbutler/tutorials/winter96/patterns/.

[67] D. L. Levine, D. C. Schmidt, "Introduction to Patterns and Frameworks," *Class Material*, Department of Computer Science, Washington University at St. Louis; http://classes.cec.wustl.edu/_cs342/.

[68] E. Gamma, R. Helm, R. Johnson, J. Vlissides, *Design Patterns, Elements of Reusable Object-Oriented Software*, Addison-Wesley, Reading, MA, 1995.

[69] B. Foote, J. Yoder, "The Selfish Class," *Third Conference of Patterns Languages of Programs (PLoP'96)*, Monticello, IL, September 1996.

[70] R. Goering, "Platform-based Design: A Choice, not a Panacea," *EETIMES*, September 2002; http://www.eetimes.com/story/OEG20020911S0061.

[71] J. Clark, *XSL Transformation (XSLT) Version 1.0*, World Wide Web Consortium, November 1999; http://www.w3.org/TR/xslt.

[72] S. Boag, D. Chamberlin, et.al., *XQuery 1.0: An XML Query Language*, Working Draft, World Wide Web Consortium, November 2002; http://www.w3.org/TR/xquery/.

[73] "XPath Tutorial," *W3Schools.com*; http://www.w3schools.com/xpath/default.asp, accessed March 2003.

[74] *Xalan Java Version 2.5.D1*, The Apache XML Project; http://xml.apache.org/xalan-j/index.html.

[75] *Xalan C++ Version 1.4*, The Apache XML Project; http://xml.apache.org/xalan-c/index.html.

[76] *Jaxen*, SourceForge.net; http://sourceforge.net/projects/jaxen/.

[77] *MSXML 3.0*, Microsoft Inc.; http://msdn.microsoft.com/xml/.

[78] M. Nic, J. Jirat, "XPath Tutorial," http://www.zvon.org/xxl/XPathTutorial/General/examples.html, accessed March 2003.

[79] *The Nexperia Platform*, Philips Semiconductors; http://www.semiconductors.philips.-com/platforms/nexperia/technology/index.html, accessed January 2003.

[80] *The Open Multimedia Applications Platform (OMAP),* Texas Instruments; http://www.omap.com, accessed March 2003.

[81] *The NiOS Embedded Processor*, Altera; http://www.altera.com/products/devices/nios/nio-index.html, accessed March 2003.

[82] *The QuickMIPS Platform*, Quicklogic Inc.; http://www.quicklogic.com/home.asp? PageID=376& sMenuID=220, accessed March 2003.

[83] *The Virtex-II Pro Platform FPGAs*, Xilinx Inc.; http://www.xilinx.com/xlnx/xil_prodcat_landing-page.jsp?title=Virtex-II+Pro+FPGAs, accessed March 2003.

[84] *Avalon Bus Specification*, Altera Inc., San Jose, CA, January 2002.

[85] *CAN Specification Version 2.0*, Robert Bosch GmbH, September 1991.

[86] *FlexRay Requirements Specification, Version 2.0.2*, FlexRay-Consortium, April 2000; http://www.flexray-group.com/.

[87] *The I^2C-Bus and How to Use It*, Philips Semiconductors, accessed 1995; http://www.semiconductors.philips.com/acrobat/ various/I2C_BUS_SPECIFICATION_1995.pdf.

[88] C. Zhang, F. Vahid, "A Power-Configurable Bus for Embedded Systems," *IEEE International Symposium on Circuits and Systems*, pp.V-809-812, May 2002.

[89] *Universal Resource Identifier (URI) Specification*, World Wide Web Consortium; http://www.w3.org/Addressing/URL/URL_TOC.html, accessed March 2003.

[90] B. Foote, J. Yoder, "Metadata and Active Object-Models," *OOPSLA '98 MetaData and Active Object-Model Workshop*, Vancouver, Canada, October 1998; http://www.laputan.org/metadata/metadata.html.

[91] P. Eles, Z. Peng, K. Kuchcinski, and A. Doboli, "System Level Hardware/Software Partitioning Based on Simulated Annealing and Tabu Search," *Design Automation for Embedded Systems*, pp. 5-32, 1997.

[92] M. S. Haworth and W. P. Birmingham, "Towards Optimal System-Level Design," *Proceedings of the 30th Design Automation Conference*, pp. 434-438, 1993.

[93] A. K. Majhi, L. M. Patnaik, and S. Raman, "A Generic Algorithm-Based Circuit Partitioner for MCMs," *Microprocessing and Microprogramming*, Vol. 41, pp. 83-96, 1995.

[94] J. Teich, T. Blickle, and L. Thiele, "An Evolutionary Approach to System-Level Synthesis," *Proceedings of the 5th International Workshop on Hardware/Software Codesign*, pp. 167-171, 1997.

[95] T. Tran, "OMAP5910 NTSC or VGA Output," *Texas Instruments' DSP/EEE Catalog, OMAP Application Report SPRA847*, June 2003; see http://www-s.ti.com/sc/psheets/spra847/spra847.pdf.

[96] K. Sabbagh, *Twenty-First Century Jet: The Making and Marketing of the Boeing 777*, Scribner, New York, 1996.

[97] A. Ramirez, et.al., "ArgoUML User Manual, A Tutorial and Reference Description of ArgoUML, Revision 0.10," *Open Publication License*, May 2002; http://argouml.tigris.org/documentation/defaulthtml/manual/.

[98] B. P. Douglass, *Real-Time UML: Developing Efficient Objects for Embedded Systems*, Second Edition, Addison-Wesley, Reading, MA, 1999.

[99] J. Warmer, and A. Kleppe, *The Object Constraint Language: Precise Modeling with UML*, Addison-Wesley, 1998.

[100] *MagicDraw™ 6.0*, No Magic Inc.; http://www.magicdraw.com/, accessed August 2003.

[101] *Poseidon for UML, Professional Edition 1.6 and Embedded Edition 1.6.1*, Gentleware AG, Hamburg, Germany; http://www.gentelware.com, accessed August 2003.

[102] *Rational Rose*™, IBM Inc.; http://www.rational.com, accessed August 2003.

[103] *Rhapsody™ in C, C++ and Ada*, I-Logix Inc.; http://www.ilogix.com/products/rhapsody/index.cfm, accessed August 2003.

[104] *Codagen Architect*, Codagen Technologies Corp., Montreal, Canada; http://www.codagen.com/products/architect/default.htm, accessed August 2003.

[105] *Enterprise Architect Ver. 3.51*, Sparx Systems, Australia; http://www.sparxsystems.com.au/ea.htm, accessed August 2003.

[106] A. Moore, *RT-Profile—A Quick Tour*, Artisan Software Inc.; http://wwwbroy.informatik.tu-muenchen.de/~rappl/RT%20Profile(new).pdf, last accessed August 2003.

[107] V. Madisetti, L. Shen, "Timing Interface Design in Core-based Systems," *IEEE Design & Test of Computers*, Vol 13, No. 4, October-December 1997.

[108] A. Burns, A. Wellings, *Real-Time Systems and Programming Languages, Third Edition*, Addison-Wesley, 2001.

[109] A. Costello, C. Truta, *CEXCEPT Exception Handling in C, Version 1.0.0*, The CEXCEPT Project, June, 2000; see http://cexcept.sourceforge.net/.

[110] P. Lee, "Exception Handling in C Programs," *Software—Practice and Experience*, Vol 13, No. 5, pp. 393-401, 1983.

[111] The Real-Time for Java™ Expert Group, *The Real-Time Specification for Java™, Version 1.0*, Addison-Wesley, 2000; see also http://www.rtj.org/rtsj-V1.0.pdf.

[112] *Draft Standard for VHDL Register Transfer Level (RTL) Synthesis*, IEEE Standard P1706.6/D6, IEEE Computer Society, May 2003; see http://vhdl.org/siwg/66_D6X.PDF.

[113] D. J. Smith, *HDL Chip Design*, Doone Publications, AL, 1996.

[114] J. Bhasker, *A VHDL Synthesis Primer*, Star Galaxy Publishing, PA, 1996.

[115] J. Bhasker, *Verilog HDL Synthesis, A Practical Primer*, Star Galaxy Publishing, PA, October 1998.

[116] D. Björklund, J. Lilius, "From UML Behavioral Descriptions to Efficient Synthesizable VHDL," *In NORCHIP02*, Copenhagen, Denmark; http://www.infa.abo.fi/~dbjorklu/publications/norchip02.pdf.

[117] W. McUmber, B. H. C. Cheng, "UML-Based Analysis of Embedded Systems Using a Mapping to VHDL," *4th IEEE High Assurance Software Engineering*, pp. 56-63, November 1999.

[118] S. J. Mellor, L. Starr "A Method for Making UML Directly Executable," *EETIMES*, November 2002; see http://www.eetimes.com/design_library/esd/dt/OEG20021115S0045.

[119] G. Martin, "Coexistence in a Multilingual Design World," *EETIMES*, June 2003; see http://www.eedesign.com/silicon/OEG20030616S0064.

[120] F. Vahid, and T. Givargis, *Embedded System Design: A Unified Hardware/Software Introduction*, John Wiley & Sons, October 2001.

[121] S. W. Grotta, "Anatomy of a Digital Camera," *ExtremeTech*, June 2001; see http://www.extremetech.com/article2/0,3973,15466,00.asp, accessed September 2003.

[122] S. W. Grotta, "Anatomy of a Digital Camera: Image Sensors," *ExtremeTech*, June 2001; see http://www.extremetech.com/article2/0,3973,15465,00.asp, accessed September 2003.

[123] I. Graham, *Object-Oriented Methods, Principles & Practice, Third Edition*, Addison-Wesley, 2001.

[124] R. Abbott, "Program Design By Informal English Descriptions," *Communications of the ACM*, Vol. 26, No. 11, pp. 882-894, 1983.

[125] R. Wirfs-Brock, B. Wilkerson, L. Wiener, *Designing Object-Oriented Software*, Prentice Hall, 1990.

[126] Z. Chen, J. Cong, Y. Fan, X. Yang, Z. Zhang, "Pilot — A Platform-Based HW/SW Synthesis System for FPSoC," *Workshop on Software Support for Reconfigurable Systems*, February 2003.

[127] C. Loeffler, A. Ligtenberg, G. Moschytz, "Practical Fast 1-D DCT Algorithms with 11 Multiplications," *Proceedings of the International Conference on Acoustics, Speech, and Signal Processing 1989 (ICASSP '89)*, pp. 988-991.

[128] W. Pennebaker, J. Mitchell, *JPEG Still Image Data Compression Standard*, Van Nostrand Reinhold, 1993.

[129] A. Messac, M. Martinez, T. Simpson, "Effective Product Family Design Using Physical Programming," *Engineering Optimization*, Vol. 34, No. 3, 2002, pp. 245-261.

[130] P. Clarke, "Gartner Trims 2003 Semiconductor Growth Forecast," *Semiconductor Business News, EETIMES*, May 2003; see http://www.eetimes.com/semi/news/OEG20030520S0009.

[131] *Functional Specification for SystemC 2.0*, Open SystemC Initiative (OSCI), April 2002; see http://www.systemc.org.

[132] *SystemC 2.0 User's Guide*, Open SystemC Initiative (OSCI), 2002; see http://www.systemc.org.

[133] G. Martin, L. Lavagno, J. Louis-Guerin, "Embedded UML: a merger of real-time UML and codesign," *CODES 2001*, Copenhagen, April 2001, pp.23-28.

[134] B. Selic, G. Gullekson, P. Ward, *Real-Time Object-Oriented Modeling*, John Wiley and Sons, New York, 1994.

[135] J. L. Diaz-Herrera, V. K. Madisetti, "Embedded Systems Product Lines," *Proceedings of Software Product Lines, 22nd International Conference on Software Engineerings*, Limerick, Ireland, June 2000, pp. 129-136.

[136] J. Withey, "Investment Analysis of Software Assets for Product Lines," *CMU/SEI-96-TR-010*, Carnegie Mellon University, Pittsburgh, PA, 1996.

[137] J. L. Diaz-Herrera, J. C. Guzman, "Product Lines in the Context of Embedded Systems," *Proceedings of Software Product Lines, 23rd International Conference on Software Engineerings*, Toronto, Canada, May 2001, pp. 19-25.

[138] M. S. Ben Romdhane, V. K. Madisetti, J. Hines, *Quick-Turnaround ASIC Design in VHDL*, Kluwer Academic Publishers, MA, June 1996.

Index